The Profitable Art and Science of Vibratrading

Non-Directional Vibrational Trading Methodologies for Consistent Profits

The Profitable Art and Science of Vibratrading

Non-Directional Vibrational Trading Methodologies for Consistent Profits

MARK ANDREW LIM

WILEY

John Wiley & Sons (Asia) Pte. Ltd.

Published in 2011 by John Wiley & Sons (Asia) Pte. Ltd.
1 Fusionopolis Walk, #07-01, Solaris South Tower, Singapore 138628

Other Wiley Editorial Offices
John Wiley & Sons, 111 River Street, Hoboken, NJ 07030, USA
John Wiley & Sons, The Atrium, Southern Gate, Chichester, West Sussex, P019 8SQ, United Kingdom
John Wiley & Sons (Canada) Ltd., 5353 Dundas Street West, Suite 400, Toronto, Ontario, M9B 6HB, Canada
John Wiley & Sons Australia Ltd., 42 McDougall Street, Milton, Queensland 4064, Australia
Wiley-VCH, Boschstrasse 12, D-69469 Weinheim, Germany

Library of Congress Cataloging-in-Publication Data
ISBN 978–0–470–82874–8 (Hardcover)
ISBN 978–0–470–82876–2 (ePDF)
ISBN 978–0–470–82875–5 (Mobi)
ISBN 978–0–470–82877–9 (ePub)

Typeset in 11/14pt, Century-Book by Thomson Digital, India

10 9 8 7 6 5 4 3 2 1

Contents

Acknowledgments

I would like to convey my heartfelt gratitude and appreciation to my parents for their constant inspiration and guidance, and to my sister, Stella, for her tireless support and assistance. I also extend my sincere thankfulness to all my friends including Radge, Matilda, and K.S. Hee for their unconditional support and friendship throughout the years. I thank Nick Wallwork, Joel Balbin, Jules Yap and everyone at John Wiley for the opportunity to share Vibratrading. I am ever grateful to Laura Paquette for her phenomenal contribution and expertise in helping me put this book together. Finally, I truly thank all my graduates for their amazing participation, insight, patience and dedication, without which this book would not be possible. In many ways, I am your student.

The deeper the vibration, the greater the creation.

Introduction

One of the key aspects of trading (and the most frustrating) is that it's impossible to predict the future. Since no trader can possess any absolute knowledge as to the future direction of price, one obvious option is to employ a "Martingale" strategy which keeps you increasing your bets until you eventually win. Unfortunately, since we don't know exactly how long any particular losing streak will last, and since most of us lack unlimited funds, this strategy is destined to fail, resulting in the total loss of our capital.

If we could work out exactly when that streak *would* end, of course, we would never lose, because we would know well in advance the precise amount of funds required to survive the streak and eventually produce a win, or a gain in capital. Even though the risk to reward ratio may be extremely low, especially on the very last bet, the trader would still come out on top.

Imagine if traders could enter the financial markets knowing exactly where the Martingale "limits" reside. Even if the price remains below the traders' entry level indefinitely, they would have the ability to coast through the losing streak to success.

I have adapted the high-risk and high-investment method of scale-trading to a safer, more powerful and adaptive tool: *Vibratrading*. Vibratrading is based on generating returns in the market from price oscillations, or vibrations. It is also implemented with reference to the concept of boundedness, which helps the trader or investor understand the type and degree of risk associated with any particular trading technique or mechanism. Trading according to the rules of boundedness is what separates vibratrading from conventional scale trading. Boundedness is all about capital preservation, which includes the strict avoidance of all capital depleting mechanisms like stop losses, long options or initiating net short positions. More specifically, boundedness is defined as the condition in which the final account equity will be equal to or greater than the initial account equity, should price retest the initial price entry level. For example, a "vibratrader" enters the market at a certain price. After a number of trades, the market returns to the initial price level. If the methodology caused equity to fall below its initial value, then that methodology is said to be unbounded. All trading strategies and mechanisms are categorized as either bounded or unbounded. The vibratrader has a choice to implement either trading mechanism within the vibrational construct, but this must be done with full understanding of the risks involved in choosing an unbounded methodology. These strategies can be used in conjunction with various diversification techniques to accomplish what most traders and investors previously thought impossible.

The Genesis of Vibratrading

Before I get into the methodology behind vibratrading, let me explain the thought process leading me to this point.

It all started with a simple question:

What if we could find a stock or instrument that would never collapse to zero, except in a total systemic meltdown of the financial market?

If we could construct or find such a fail-safe investment, then we would only need to preserve our invested or traded capital in the market long enough to either:

1. experience a favorable upside move, or
2. accrue enough returns to reduce our overall cost basis.

To eventually profit and avoid capital erosion, we must stick to methods which build capital and avoid all capital depleting activities and mechanisms. Of these mechanisms, one is the most detrimental. To keep our capital intact, we must avoid using the single most capital-depleting mechanism ever created, ironically called the stop loss. When investors have a working order to sell at a certain price, regardless of circumstances, they may protect themselves from loss but also risk gradual depletion of capital funds. If the position is repeatedly stopped out, the account equity will inevitably be wiped out.

If we avoid the deployment of all stop loss mechanisms in our trading and investing activities, we will need sufficient capital to sustain a fall in price down to zero for longs (investments made with the hope of prices going up). At the same time, we need enough capital to hold the shorts: stocks sold with the hope that the price will go down, thus allowing us to repurchase and make a profit. Since there is no upper boundary to price, we would need essentially unlimited capital to hold short losses, therefore the only feasible solution is to initiate and hold longs. As the maximum that we can lose is the amount that we paid for shares, the limited risk involved is capped.

Accordingly, we should only go long for vibratrading to work. In other words, traders must buy low and sell higher, and never sell high with the intention of buying back lower. As shown above, shorting would expose the trader to unlimited upside losses, especially without the protection of a stop loss mechanism. To protect the capital from depleting permanently over time, that is, to ensure that trading capital is always "bounded," all trades must close in profit, as opposed to a loss. Wouldn't that mean that all long exits would be profitable? Well, upon deep reflection, yes!

On the other hand, this does not mean we should keep buying as prices rise. There is no upside limit to price and therefore we would need unlimited capital to keep on buying as the price rises. Some think we could always just buy on profit. That is one option, but we need to be prepared should price decline prior to the opportunity to buy on profit. Remember, we cannot use a stop loss. If all capital has already been allocated to that one long position, then there is little recourse except to hold that position until price returns to the original entry level. If this should happen to a contract for difference (CFD) trader without the capital to sustain the long position all the way down, then positions could be liquidated for violating the minimum margin percentage level.

So does this mean we should keep buying as price falls? Realistically we could, provided we have enough funds to support our downside buying expedition. By buying more at a lower price, we are averaging down and reducing the average cost per unit of the investment. The difference when averaging down with longs, as opposed to averaging up with shorts, is you at least have an idea of the maximum capital you need to maintain your long positions down to zero, or what I call Zero Test Point. There is no equivalent cap when averaging up with shorts. Therefore, if you plan well in advance, the idea of averaging down is totally workable.

So far, it looks like even if we avoid all capital depleting mechanisms, we may still be holding on to non-performing long positions. However, what if price continues to oscillate at the bottom of the market? In that case, we could continue buying low and selling higher every time price vibrates, and in the process extract profit with every vibration. That would mean that we could continue to generate returns indefinitely. Of course there must be a loophole somewhere; for example, what if the shares of the stock plummet to near zero?

Actually, our oscillation profits will be high in that situation, as we can buy and sell even more shares due to the lower price. We would, in fact, make greater vibrational returns at lower share prices. We take advantage of this price leveraging effect.

If the stock plummets directly to zero and winds up, however, there is no way to profit—unless there is virtually no possibility of the stock or instrument ever testing zero. Fortunately, there are three main categories of instruments that never reach zero (except under exceptionally adverse market conditions).

The first is called an Exchange Traded Fund or ETF. An ETF is a stock with a share price based on a basket of component stocks. ETFs have a built-in replacement mechanism that automatically replaces any stock that is failing to meet the fund's criteria for inclusion within its large basket of component stocks. As such, the share price of an equity-based ETF can only hit rock bottom if every stock in the basket plummets and collapses to zero simultaneously. That scenario is incredibly improbable—it would be financial Armageddon!

The second category is Index Future Contracts, which involve an agreement to sell at a certain price at an agreed future date. As with ETFs, index future contracts are based on a basket of component stocks and as such will never test zero except in exceptionally adverse markets. Index investors receive a separate return, the roll yield, when they roll trades over periodically. That return is negative when far out futures prices are higher, or in "contango." Unfortunately, we cannot employ index future contracts in vibratrading due to the effects of negative roll yield.

The final category is commodities. It is virtually impossible for a commodity that is not financial-based to hit zero test point. When was the last time we saw precious and base metals, oil, wheat, corn, or sugar at rock bottom? Never! But we can only vibratrade the spot prices of these commodities via CFD platforms. We cannot use futures contracts to gain exposure to these commodities, due to the negative roll yield mentioned above. We also cannot use long options to trade or invest in these markets; it could expose the trading account to capital unboundedness. As a result, we will only focus on ETFs and CFDs in vibratrading. As a rule, we do not vibratrade single stocks as there is no mechanism to prevent them from winding up. But I will introduce a vibrational technique called Oscillatory Propagation should a stubborn vibratrader insist on trading single stocks.

Okay, things are starting to look much better. But if the price of the equity-based ETF stock just stays completely still and "flatlines," it appears there is no way to profit within the system. In reality, all we have to do is incorporate options into our vibrational construct to generate profit. We never buy options, as they deplete capital if you fail to achieve at least a breakeven trade. This was the very reason we avoided using stop losses. In flatlining markets, we sell options instead.

Yes, we all have heard of traders and investors losing the shirts off their backs from trading short options. But, if we deploy short options within an oscillatory or vibrational trading construct, then those positions will be completely riskless, unless the price of the stock or instrument falls to zero. The remainder of this book will show you how to maintain short options without risk, while your predetermined working capital continues to generate returns indefinitely.

I will start by teaching you some of the basic knowledge required to fully grasp vibrational trading systems. You will then learn to set up, construct, and implement some of the most effective bounded and unbounded trading methodologies for generating consistent (as well as exponential) returns in the markets. You will also learn to never fear falling markets again, and in fact, to look forward to such bear action! Buy and hold is comparably inefficient as vibratraders continue to generate returns instead of passively hoping for price to return to previous levels. Eventually, we will see many examples of bounded and unbounded vibrational constructs for generating consistent income in all manner of markets, taking conventional scale-trading and averaging strategies to the next level.

LAYOUT

This section describes the basic layout of the book on a chapter by chapter basis. Readers are advised to start from the beginning, going through all the chapters sequentially. Various terms, concepts, and techniques will be introduced systematically.

Chapter 1 briefly describes the nature of vibrational strategies and techniques for extracting profits from the markets. It discusses the difference between directional and vibrational trading. There is also a brief explanation of the concept of boundedness.

Chapter 2 covers the basics of order entry with a look at the various characteristics and functionality of trade orders. Special emphasis is placed on those orders that are used in vibrational trading. Besides covering some terminology and trading definitions, this chapter also delves into the various price- and time-triggered orders and includes a very useful graphical representation for easy referencing.

Chapter 3 reveals the objectives of vibratrading, including employing it as an income strategy. This is followed by a basic introduction to the concepts and definitions applicable to vibrational trading.

Chapter 4 goes over the control of risk in vibratrading, covering the various types of risk and its control mechanisms, as well as topics on diversification, hedging, long options, and the important Disposable Capital Rule.

Chapter 5 introduces the mechanics of equity price action which represent the "nuts and bolts" required to fully comprehend the various scaling constructs introduced in later chapters. It explains how to calculate simple profit and loss, market value of total investments, and various leverage factors. Important concepts like price, buy, and money leverage ratio are of particular relevance with regard to the selection process of a suitable ETF or CFD, as well as the vibrational construct desired.

Chapter 6 includes the analysis of price action itself, with regard to single and multiple hedged and unhedged positions, with simple calculations of hedged or soft-locked profit and losses. The concept of average price, break-even point, and net position is thoroughly treated. The implications of negative spread bias are also highlighted.

Chapter 7 presents the mechanics and dynamical aspects of "boundedness," a unique concept that gives rise to the rules of vibrational trading. Boundedness originates from the important role of capital preservation in trading and investing, and includes three aspects: Range, Directional and Order Entry. Boundedness not only dictates how the scaling mechanism and constructs should work, but also indicates which trading strategies and techniques are bounded or unbounded.

Chapter 8 is critical as it explains the construction of all the scaling mechanisms of the vibrational methodology. It starts by describing the general properties of the pyramidal vibrational structure with references to scaling factor, foundational stability, conversions, and the Buy-Sell mechanism. All this is followed by a detailed description of the main scaling mechanisms used in vibratrading.

Chapter 9 explains the mechanisms and methodologies under the Pyramidal Based structure and vibrational elements like termination and share multiples, along with its constructs: the Null, Profit, and Phi-Based constructs. Vibrational hedging follows with illustrations of the numerous forms of hedging for long and short vibrational returns, with special focus on the Short Scaler and Rider techniques.

Chapter 10 describes in detail the diversification strategies and levels used in vibrational trading. In fact, these levels of diversification lay the foundation for a virtually indestructible portfolio by overcoming the effects of both systematic and specific risk. The idea of the pyramidal "floor" is examined, followed by an illustration of the vibrational constructs. Diversification is treated in detail with reference to its six levels. Particular attention is paid to the fifth level of diversification: Oscillatory Propagation. Finally, the difference between market-driven and structurally-driven correlation is presented along with various charts depicting its behavior.

Chapter 11 explains the difference between timing the market for direction and timing for volatility. A range of volatility techniques are described specifically for use within the vibrational constructs for added profit potential, which includes event trading, range zoning, and effective range scaling.

Chapter 12 represents the culmination of all vibratory and trend capture techniques. The reader is taken through the practical steps of selecting and putting together bounded and unbounded vibrational, bi-directional, and directional constructs. It also covers cost reduction techniques. A thorough analysis of the operational and functional characteristics of the vibrational grids is carried out. The bounded and unbounded trend capture constructs are then introduced, with examples using CFD traded commodities. Numerous other vibrational techniques like Macrosiso Vibrahedging, Martingaling, and Zero Cost Hedging are then introduced. Finally the vibratrader is shown how to generate consistent returns via the use of the average period range.

Chapter 13 explains the many aspects and characteristics of ETF behavior, with emphasis on its inherent risks, advantages, diversification, and leveraging. There are numerous examples with charts showing the differences between market- and structurally-driven correlation. Sector rotation and basic intermarket analysis is covered. Non-linear performing

ETFs and ETNs are examined along with asymmetric leverage, negative expectancy, and the daily reset feature. The fifth level diversification strategy is also discussed within the context of the replicated and hybrid portfolio.

In **Chapter 14,** vibratrading is compared and contrasted with some of the more popular trading and investing systems, highlighting its overall efficiency. This section examines how vibratrading overcomes many common shortcomings, and will be of particular interest to those who are currently trading and investing with these popular systems. Vibratrading presents various ways to augment and improve such systems with a view to minimize risk and maximize profitability.

In **Chapter 15**, some simple options techniques are introduced to supercharge the vibrational constructs for increased level and consistency of returns for both volatile and flat line markets. It examines the use of riskless short options as an entry mechanism for vibrational trading. Covered calls are also discussed as part of a "covered short strangle" strategy for maximizing profitability.

Chapter 16 presents concluding remarks and a summary of the purpose, use, and future of vibratrading.

CHAPTER 1

Challenges to Conventional Trading and Investing

In this opening chapter I will explore some of the common issues faced by conventional traders and we will begin to discover the vibratrading difference and advantage.

DIRECTIONAL VS. NON-DIRECTIONAL METHODOLOGY

By far, the most popular trading approach is directional. This requires a trader to make a bet on the future direction of price. We see this approach used heavily in scalping, day, and swing trading, and especially in instruments offering medium to high leverage such as Foreign Exchange (FOREX), commodity futures, CFDs, and options. If price moves favorably, all is well. Problems surface any time price moves adversely, resulting in capital erosion by the amount risked on each trade. Therefore, to avoid blowing out the entire account, a directional trader must ensure that the system is profitable indefinitely, or until the trader decides to cease all trading activities. If a trader fails to maintain this ongoing condition of consistent profitability, all trading will eventually come to an end, without any means of extracting further profits without injecting new capital.

Furthermore, once the trading account's drawdown (percentage between the equity peak and the trough during a specific period) is over fifty percent, recovery is extremely difficult. If trading is allowed to continue, it is very likely that all capital will be depleted in the process.

To overcome this, many traders resort to non-directional strategies through which they hope to negate the effects of directional risk by attempting to profit bi-directionally. There are basically two ways that traders can accomplish this feat.

A trader may employ a "straddle-type" breakout strategy. Two stop entry orders (orders to enter the market at a less favorable price) are made in an attempt to encapsulate the market, hoping for a breakout in either direction. Unfortunately, many traders experience very rapid oscillation losses across the straddle zone, especially with false breakouts in a prolonged inactive, or sideways, market. Even if the initial breakout was successful, trying to gauge a suitable exit becomes the new challenge. If the exit is taken too soon, there may be insufficient profit to offset past and future oscillation losses. On the other hand, if there are already some oscillation losses, exiting before breakeven will lock them in. While the straddle breakout may seem to overcome the directional risk issue, it comes at a high cost when the market reverts to sideways or volatile behavior.

Another popular way to eliminate directional risk is to resort to so-called "non-directional" options strategies. The main problem is that those strategies must be directional to some degree, as they require the trader to predict the degree and time of arrival of future price movement or reaction. It still requires price to respond in a particular manner, and accordingly, the trader ends up having to predict the right strategy to use in order to collect positive premium and keep it.

For example, a long straddle option strategy requires price to move at least a certain distance before it is profitable, even though it is already "in the money" (ITM), or worth exercising. The challenge of breaking even is further exacerbated if the long straddle was purchased during a period of high implied volatility, which increases the premium of the long options. Hence you need a certain amount of price excursion to overcome the premiums of the long call and put options in order to profit. If price fails to move, you lose both premiums and experience capital erosion. Conversely, in a non-option-based bidirectional breakout, you would incur no cost whatsoever for unfilled pending stop entry orders on either side of the bidirectional breakout. In the short straddle option strategy, price has to be stationary, or remain within a specific range. If price makes a significant excursion in either direction, the trader will lose and capital will be depleted in the process.

Unlike the conditional terms of straddle options, vibratrading allows the trader to use the very same entry strategy for all price and market outcomes. The vibrational constructs and mechanisms will generate returns in up, down, and sideways markets, including perfectly flatlined markets, via the implementation of riskless short options positions that do not deplete capital.

In other words, vibratraders do not need to predict the future outcome of price or which strategy to use. All vibrational entry mechanisms extract profit, no matter what. This means that if price falls after a long entry, you will still generate vibrational returns, even if price never returns to its original entry level. In fact, you will be exposed to the greatest returns when the market is at its weakest.

On the other hand, should price explode to the upside, the very same scaling mechanism will capture trend profits. More positions will be pyramided in to maximize profit potential once certain conditions are met.

We will learn how to extract more vibrational and trend based returns without added risk or capital.

The reason that these unique vibrational and trend-based mechanisms can capture profit in all markets, regardless of direction, is their ability to adapt to whatever scaling or trending construct is required for that market. This ability to capture bounded returns is only possible if entry was executed at or below the pyramidal apex level. The apex represents the highest point in a pyramidal structure where long entries are still bounded. Entering long positions above the apex leaves insufficient capital to hold these positions down to zero, thereby risking a margin call. This is especially relevant when trading with high leverage. In a "cash" account, all shares are bought and paid for up front and the concept of an apex is redundant.

Return derived from entries above the apex is unbounded in nature. In fact, every bounded vibrational entry must be at or below the apex level in order to accommodate every possible future price action. This important bounded entry condition will be covered in subsequent chapters.

PROBLEM OF MAINTAINING LONG-TERM CONSISTENT POSITIVE EXPECTANCY

To be profitable in trading, you will need to use either a Martingale-type system with pre-knowledge of the maximum number of losing trades possible, or an anti-Martingale system. That system only increases successive bets if it is profitable, and in the process must able to maintain a consistently positive trade expectancy or average return. For those unfamiliar with the concept of trade expectancy, it simply refers to the profitability of a series of trades in terms of winning percentage to the average dollar win or loss per trade. Basically, expectancy tells you whether you have been profitable or otherwise, over a number of trades, based on your entry and exit rules.

Every trader knows, or rather should know, that the only factor you can actually control is your dollar win or loss per trade, also referred to as the reward to risk ratio. You have absolute control over your entries and exits, and as such, you have total control over the amount of profit or loss per entry. But you have no control whatsoever over the number of wins or losses. Winning percentage is purely determined by the markets, and you cannot influence the future direction of price. No amount of fundamental, technical, statistical, or behavioral analysis (no matter how advanced or sophisticated) can possibly forecast with perfect accuracy the subsequent direction of price. To make matters worse, expectancy is just that—it "expects" the win-loss ratio to remain constant. As we well know, that is akin to wishful thinking. The markets are under no obligation to generate a consistent winning percentage for anyone.

This unrealistic attempt to keep a trading system's expectancy positive and consistent indefinitely is one of the biggest challenges in directional trading.

> *Not only does the directional trader have no control over the win-loss ratio, the directional trader also has no control whatsoever over the win-loss distribution!*

This results in what I call the "Expectancy Box" problem. I conduct an entire masterclass in "Stochastic Maximization" to help traders overcome the issues of extracting profit from randomness. This does not mean that directional trading, which I define as trading with a stop loss mechanism, is bad in any sense. What I am saying is that it is tremendously difficult to maintain consistent profit indefinitely.

Nevertheless, the concept of expectancy remains important to directional traders, despite being highly difficult to sustain, especially in the face of inexplicable market expression. The challenge of maintaining consistent positive expectancy, coupled with the uncertainty of predicting future price direction, is often considered a cause of traders losing their account completely, which happens more often than we'd like to believe. In vibratrading, we do not need to predict price action at all; it can only add to the effectiveness of the methodology.

Even gamblers can be profitable over the very short term. Though the application of money management plays a pivotal role in the final determination of trading expectancy, the actual task of producing net profits is still formidable, if not near impossible, for most retail traders and average investors. Profitability under the vibratrading methodology does not depend on the win-loss ratio—since every trade only closes in profit there is no win-loss ratio. Therefore, vibratrading, be it bounded or otherwise, totally circumvents the sheer difficulty of maintaining a consistent positive expectancy in the long term. The concept of winning percentage and reward to risk ratio is not applicable.

PREDICTIVE VS. REACTIVE APPROACHES TO RISK IN TRADING

Another issue that plagues traders is the use of price prediction as part of a trading strategy. It is universally accepted that price and market action are random or at best, semi-random in nature. You cannot access all relevant trade information instantaneously to assess the actual forces of supply and demand. Even if you could, you would still require information as to the possible future actions of all market participants in order to anticipate the supply and demand at approaching price levels. You would also need to be able to anticipate unexpected market "shocks," or catalyzing events, including the range of possible reactions of all participants. Not only do you need to know of any possible current or future events, you also need to consider the tenuous intermarket interactions that may affect your trades or investments. As you can see, maintaining a reasonable and consistent winning percentage while maximizing the reward to risk ratio is going to be a never-ending challenge to the directional, predictive trader.

The reactive trader faces similar challenges, the difference being the reactive trader only initiates a trade entry after price has confirmed a breakout, pullback (a rising of price from its bottom), or throwback (a falling of price from its peak). For example, a reactive trader would enter a pullback from a support level via a buy stop entry order, after price has confirmed the pullback, whereas the predictive trader would enter at the same support via a buy limit entry order, before any proof of a pullback is evident. Reactive traders may also try to predict future price direction, but they will only initiate a position once price has confirmed a favorable move that was predicted by the analysis. Therefore the reactive trader is consistently late in initiating entries, whereas the predictive trader enters a position in anticipation of its outcome. In terms of risk profiling, the predictive trader may be "aggressive in time" but "conservative in price," whereas the reactive trader is "aggressive in price" but "conservative in time." This is the dualistic nature of trader risk profiling. "Aggressive in time" means to initiate a position before price confirms a favorable or desired move, whereas "conservative in time" means to initiate a position after that has occurred. Someone who is aggressive in price will initiate a position at a less favorable price level; conservatives in price initiate a position at the most advantageous price level.

As mentioned, the concept of risk and reward is not at all applicable to vibratrading. The basic idea of return on investment (ROI) applies, but the returns are also defined over time, in addition to a fixed initial invested value.

Also, the relationship of time and risk in vibratrading totally contrasts with traditional directional trading. Generally, the more time you spend in the markets, the lower your chances of achieving consistent profitability. But in vibratrading, the more time spent in the market, the less the risk. This is because the investment's cost basis is constantly being reduced via vibrational returns. Therefore it is advantageous to remain in the market as long as possible, preferably indefinitely.

TRADER INACTIVITY AND VOLATILE PRICE ACTIVITY

In most directional trading approaches, a system will suffer losses in the form of slippage when a trade entry or exit is executed at a price that is different from the expected price. These losses only occur with stop orders. Orders being filled at an unexpected price above a buy stop will result in additional losses, called negative slippage. Negative slippage results in reduced profits for those entering long positions and affects sell stop orders. Either way, stop entry or exit orders may have a negative effect on trades. In an unbounded vibrational construct, the vibratrader may employ long and short stop orders for entry and exit, be it pending or instantaneous.

But with limit orders, slippage can only have a positive effect. Being filled at a price below a buy limit order may result in additional "potential" profit for those entering long positions, while ensuring additional real profit for those exiting short trades. In other words, limit orders may experience positive slippage, but never negative slippage.

So, in bounded vibratrading, only limit orders are used. The principle of boundedness forbids the use of stop entry or exit orders in order to prevent negative slippage and, ultimately, capital depletion. This means that, if boundedness is to be maintained, all entries and exits should be executed strictly via buy and sell limit orders respectively. In unbounded vibratrading, we employ stop orders, but we must understand and accept the risk of capital erosion that may arise from negative slippage and price gapping in a volatile market. In bounded vibratrading, we welcome market volatility because trading with limit orders will only result in positive slippage, greater profits, or more advantageous entries and exits.

Should a vibrational trader experience price gapping over any limit orders placed for exiting a position, the trader will experience additional profit. On the other hand, should that trader experience price gapping over any limit orders placed for entry into a position, the trader may experience additional "potential" profits upon a favorable exit, since the trader has entered at a much lower entry price for longs. This represents a real trading and investing advantage, or edge.

Hence, all price or market volatility, whether the result of some economic release like earnings or a major broad market or inflation report, will always be advantageous for bounded vibratraders. They either exit with added profit in their long positions, or enter into positions with a favorably lower buy price. It's a win-win situation.

Finally, with limit orders, there is no missed opportunity. A vibrational trader can profit from certain inaction. For example, if a vibratrader misses an exit point by forgetting to place a sell limit order, he or she may experience additional profit if price continues to proceed favorably. Similarly, vibrational traders or investors who miss an entry point by forgetting to place a buy limit order may experience additional potential profit upon a favorable exit, since they bought at a lower price. The lower we buy and the higher we sell, the more we make. Generally, we always buy as low as we reasonably can and sell as high as we possibly can. Unfortunately, inaction in directional trading may prove disadvantageous.

SUBJECTIVITY VS. OBJECTIVITY IN TRADING AND INVESTING

Another challenge in trading or investment analysis is the issue of subjectivity. There is subjectivity in trade execution as well as the analysis. There are countless ways of analyzing price and market action. Fundamental, classical, technical, neural, statistical, and behavioral are but a few of the more popular forms of trade analysis used today. The main challenge in all forms of trade analysis lies with the subjective nature of its interpretation. This also applies to the subjective nature of trade implementation, that is, process of selecting a particular trade execution technique.

The task a trader faces is trying to match up a single form of price action analysis, which is subjectively interpreted, to one form of trade execution, which is subjectively selected and implemented. That daunting effort is intended to create a trading system which accurately anticipates the action of price and market reality.

Fundamental analysis is the study of intrinsic value, including cause and effect relationships between geopolitical and economic macro- and micro-infrastructures. However, there is no absolute way to verify, or even derive with any certainty, future market action from such a complex web of economic, political, financial, and geographical factors. The sheer amount of variability and modes of interaction inherent in each factor renders the final interpretation subjective. Nevertheless, there is some fundamental persistence of trend or volatility action, especially in the long term, where global fundamentals tend to exhibit more accuracy. Short term risk, on the other hand, is not commensurate with larger market fluctuations.

Classical technical analysis also suffers from the effects of subjectivity, namely interpretation and application to trade execution. For example, even the drawing of a simple trend line can be highly subjective due to the many ways it can be defined. Once a preferred definition of how a trend line is to be constructed is decided upon, the preferred way in which to trade this subjectively defined trend line must laid out as well. Do we trade the breakout or the pullback at the trend line? Do we use a time, percentage penetration, intraday, or pullback trigger? Price action violating the trend line creates a trigger. Depending on the type of filter we use to define a trade entry, the fulfillment of those filter conditions results in the triggering of a trade. As you can see, both the signal and the trigger are subjective choices. Whether price will actually react favorably and consistently with the entire signal-trigger-filter set-up is undetermined, especially in the long term.

The vibrational construct, conversely, is objective, in the sense that the entry and exit levels may be positioned anywhere within the pyramidal structure and be able to extract vibrational returns. There are no fixed rules for determining price levels for entry or exits, though there are various vibrational scaling constructs that can easily suit the historical price volatility of the particular ETF or commodity that you are trading on the CFD platform. It is "subjective" in the sense that you still have choice in the type of vibrational scaling construct you build to match a stock's historical volatility, but your choice is never erroneous, since scaling constructs can only determine the speed of your returns. There is never a "wrong" entry in vibrational trading as the entry mechanism can be easily adapted and morphed to accommodate any market or price action. A vibratrader can enter long at any level, even modifying subsequent levels while retaining full trading flexibility, as long as the total allocated capital is conserved for that particular vibrational or scaling construct.

FILTERING AND TRADE SIGNALS

Vibratrading does not require any form of trade signals. The filtering and triggering is straightforward. In fact, no filtering is required to enter or exit a trade in vibratrading, as all entries and exits are triggered at the predetermined grids or price levels particular to that vibrational construct. Pending orders allow the long entries to be filled at the vibrational grid's ask price and long exits to be executed at the bid price. Trading cost, in the form of the bid-ask spread, is not visible on the final Profit and Loss statement (P/L) because it has been "absorbed" by the amount of work required to produce a profit. This will be covered later under Negative Spread Bias.

In summary, vibrational entries and exits are totally modifiable at any time, and may be aligned to match the stock's historical volatility and price action behavior for more expedient returns. In essence, there is never a wrong entry or profit exit in vibratrading, as "wrong" only implies a less expedient rate of return. Therefore there is no "loss," per se, in the vibratrading methodology, and returns are defined primarily as income over time from vibrational returns.

In vibratrading, just as income is derived from vibrational returns, capital gain is derived from trend profits accumulated in the construct. It is akin to owning a house where you collect income in the form of rental payment and experience capital gain as the house price rises. Income, the "rental" return, helps reduce the cost basis of the house if it is not spent and accrues. On the other hand, income can be spent, as it does not affect the integrity of the house in experiencing capital gain in the future. Hence we may spend or exhaust vibrational income as we please, as the vibrational construct represents a money-making machine that, ideally, churns out a regular income. Spending income does not affect boundedness (with the exception of "scale to breakeven exits," as you will learn later). Capital gain is of course desired but is not necessary in order to survive as a vibratrader. We do not need our house price to rise in order to make money. But we must never intentionally allow our vibrational construct to experience capital depletion by closing trades with loss. In vibratrading, paper loss, or drawdown, is not permanent loss, as the full earning potential is preserved or bounded. In fact, drawdown is usually required in order for the vibrational construct to work efficiently. This is only possible as we are trading an instrument that virtually cannot test zero. Unfortunately, paper loss is real loss in directional trading. Though it is a fine distinction, it is of utmost importance.

To sum up—Vibratrading:

- Makes scale trading affordable in the equity and commodity based markets via ETFs and CFDs. It used to be available only to institutional traders or "big boys" with access to the funds required to exploit this type of methodology in the commodity futures market.
- Instructs traders and investors, for the first time, on how to implement a diversified, grid-based, scale trading methodology in the equity and commodity based markets with a clear, easy to implement plan of action.
- Presents the most advanced and comprehensive treatment available on the art and science of scale trading and price averaging.
- Makes the new techniques of vibrational trend profit capture accessible to the average equity- and commodity-based trader and investor.
- Supersedes conventional scale trading techniques by incorporating specific techniques for speed and consistency of returns.
- Shows traders the true meaning of non-directionality, and how to consistently extract profit, irrespective of market direction, especially with the implementation of vibrational mechanisms and short options.
- Uses the exact same vibrational scaling entry mechanisms to extract profit regardless of market direction. That is, you do not need to predict

the use of the right entry strategy for an expected market outcome, unlike supposedly "non-directional" options strategies.

- Can be adapted for FOREX and futures trading.
- Teaches the important techniques of securitizing and monetizing profits with emphasis on risk-free vibrational share accumulation, which also functions as a false breakout protective mechanism.
- Overcomes the weaknesses inherent in conventional futures scale trading, dollar cost averaging, and value averaging.
- Uses the principle of "boundedness," which protects the trader's true profit potential regardless of trade performance or drawdown.
- Is totally objective in nature and hence avoids the challenges associated with subjective interpretation of price and market action.
- Eliminates the need to "time" the markets as entry can be initiated anytime.
- Demonstrates how a scale trader can enter the market at any price level, not only the conventional lower end of the instrument's historical range.
- Requires no prior knowledge of fundamental or technical analysis.
- Presents a unique perspective by defining returns in terms of time rather than the conventional approach of defining returns based on initial investment
- Involves both trend trading and counter-trend trading, and works simultaneously as a short and long term methodology.
- Is fully customizable; there is no need to adhere to any preset scaling constructs.
- Profits in all market environments, bull and bear, and makes significant returns when the market is at its weakest as well as when it is rising rapidly.
- Implements riskless short options to profit from total price inactivity.
- Introduces various levels of diversification for maintaining the overall integrity of the vibrational scaling structure as well as overcoming "systemic risk."

CHAPTER 2

Understanding the Basics of Order Entry

In order to participate in the markets, we must place entry and exit orders. This section will briefly cover some of the more common trading definitions and types of orders available.

COMMON TRADING TERMINOLOGY AND DEFINITIONS

Going Long vs. Going Short

To go long means to enter the market with a buy order, with the intention of buying low and subsequently selling higher. This is also called a buy to open order. A sell to close order is used to exit a long position, also called liquidating a long position. Every long entry will exit with a sell order, be it a profit or loss exit.

To go short, on the other hand, means to enter the market with a sell order, with the intention of selling high and subsequently buying back lower. As you may expect, this is referred to as a sell to open order. A buy to close order is used to exit or cover a short entry. In summary, to exit longs, we liquidate, and to exit shorts, we cover.

A trader is said to be square when he or she has no directional bias in the market. Being square means having either no positions in the market (in cash), or perfectly hedged positions in the market.

Perfectly hedged describes a trader without net long or short positions. Net long means your overall positions consist of more long positions than shorts, and vice versa. Cash position, or being in cash, means having no positions, and hence no rollover interest or spreads.

Bid and Ask Prices

To open a long or cover a short position, you place an ask price with a broker. An ask price is the price at which the market sells to you. The bid price is the price that you place with a broker to sell to open or sell to close. The bid price is the price that the market buys a stock or a position from you. The difference between the bid and ask prices is called the bid-ask spread. This represents a trading cost, in addition to commissions, negative slippage, interest charges, and so on. At the market means the currently traded price of the stock or the underlying instrument.

Price Triggered Orders

Price triggered orders are activated once a certain price target is hit. These orders are generally in the form of pending orders or working orders. Instructions may also be left with the broker to enter at a specified price level, which the broker may execute in the form of a market or limit order.

Market Orders

Market orders essentially instruct a broker to buy now at any price, which is also referred to as "buying at the market." The main focus is on getting into the market, rather than on price. Market orders normally assure entry or exit, with the advantage of faster fills. The downside is usually receiving a bad price fill, due to negative slippage. Some market orders and price triggered orders represent the kinds of stop orders not used in bounded vibratrading. We do, however, employ buy or sell stop orders in unbounded vibratrading, especially to catch a trend as in the Macrosimo and Upside Macrosiso scaling mechanism.

COMMON ORDERS

- **Buy Stop Orders** are instructions to buy once the market hits your intended price target. They are used for entries and stop loss exits. For entries, you want to buy high and sell even higher, so you expect to trade through the intended price level to the upside. Buy stop orders are also used to cover shorts, when they act as a stop loss order. You are not permitted to place buy stops below the market. To buy below at any available price, you may use a Market If Touched (MIT) buy order, which is activated once the stock falls to a certain level. Here, you expect price to fall first before rebounding.
- **Sell Stop Orders** instruct a broker to sell at any price below the market once it hits your target price, and are used for entries and stop loss exits. For entries, a trader sells low to cover even lower. You expect the price to trade through your price level to the downside. Sell stops act as stop loss orders for exits when a trader liquidates long positions. Just as buy stop orders are not permitted below, sell stops are not permitted above the market. To sell above the market, a Market if Touched (MIT) sell order is used, which is a market order that is activated once the stock rises to a certain level above the market, expecting price to rise first before reversing.
- **Limit Orders** essentially instruct a broker to enter the market at a predetermined price, or better. Limit orders have the disadvantage of fills that are not guaranteed. The upside is that you will not get filled at a price that is worse than instructed. All slippage via limit orders is profitable (also called positive slippage). There are several kinds of limit orders, which are in turn divided into sub-categories.
- **Buy Limit Orders** are instructions to buy at a specified price (or lower) below the market. They are used for entry and profit exits. For entries, the intention is to buy low and sell higher, and you expect price to rebound to the upside. For exits, it represents profit taking on the shorts. If you place buy limits above the market, you will be filled instantly at market, since it is the lower or better price. An MIT buy order, as mentioned above, is sometimes used instead of a buy limit order as there is no guarantee that a buy limit order will be filled.

- **Sell Limit Orders** have brokers sell at a specified price or higher above the market, once it hits your target price, and are used for entry and profit exits. For entries, the intention is to sell high and buy back lower, as you expect a price throwback from the upside. For exits, it represents profit taking on the longs. It is possible to place sell limits below the market but if you do, you will be filled instantly at market, since that is the higher price. Sell limit orders may be used above the market but there is no guarantee of being filled. For a better chance of being filled, an MIT sell order should be used.
- **Stop Limit Orders** instruct brokers to enter the market once a predetermined price is hit, but not beyond a limit price. If price triggers the market order to execute a buy or sell, but has overshot the limit price before execution of the buy or sell, it will not be filled. This is the reason why it should not be used as a stop loss order. There are two main kinds of stop limit orders.
  - **Buy Stop Limit Orders** are instructions to start buying from a specified price but not beyond the set limit price. They are used for entries only, where the intention is to buy high but not any higher, and you expect price to rise.
  - **Sell Stop Limit Orders** are instructions to start selling from a specified price but not beyond the set limit price. They are used for entries only. For entries, the intention is to sell low but not any lower, as you expect price to fall but you do not want to sell too low.
- **Time Triggered Orders** are activated once a certain time has elapsed. Most time triggered orders are market orders, and all can be used for entry or exit.
- **Market On Close Orders** (MOC) are executed during the last minutes of the trading day by the brokers. MOC orders to buy or sell are executed at or near the closing price.
- **Market On Open Orders** (MOO), on the other hand, are orders to buy or sell at the market's opening price, or near to it. MOO orders are executed at the market open.
- **Good till Cancelled** (GTC) orders instruct the brokers to keep an order active until communication asking otherwise.
- **Day Orders** instruct the brokers to keep an order pending only until the end of the trading day.

ENTRY ORDERS FOR BOUNDED VIBRATIONAL TRADING

We will use only limit orders, both for entries and exits, in bounded vibrational trading. Scaling mechanisms, including the Bounded Microsiso, Bounded Macrosiso, Bounded Macrososi, Bounded Upside Macrosiso, and the Microsiso mechanism will use only buy limit entry orders and sell limit orders for profit exits, in accordance to the rules of Order Entry Boundedness. All unbounded scaling mechanisms like the Unbounded Microsimo and the Unbounded Upside Macrosiso use stop entry and exit orders. They violate order entry boundedness.

Figure 2.1 displays the range of commonly used entry orders, including MIT orders, which may be used for entry and exit.

Figure 2.2 displays the range of commonly used exit orders.

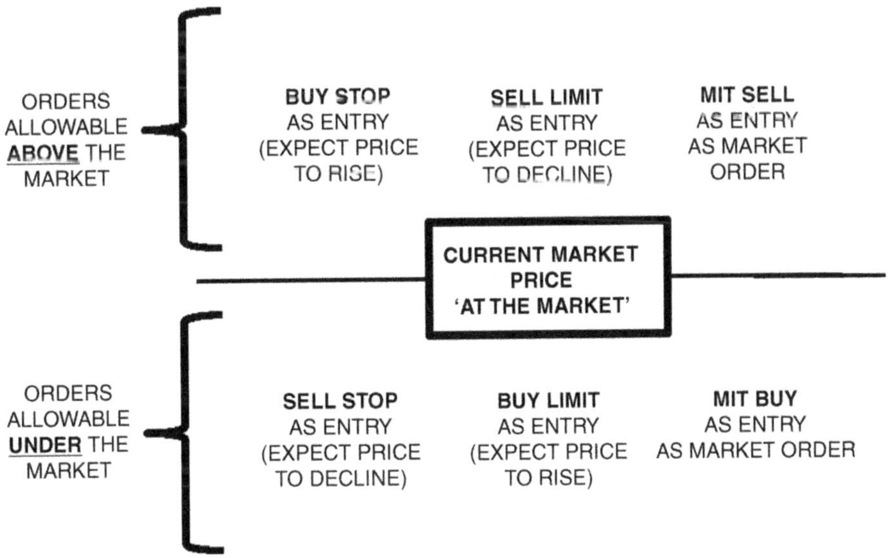

FIGURE 2.1 Graphical Representation of Entry Orders

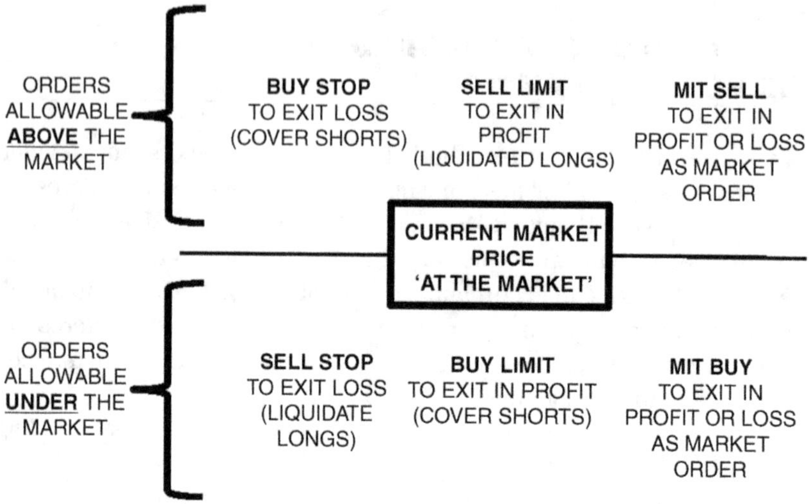

FIGURE 2.2 Graphical Representation of Exit Orders

CHAPTER 3

The Objectives of Vibratrading

Vibration trading, or vibratrading, is the extraction of oscillation profits resulting from price fluctuations. In essence, it is the conversion of vibrational energy into fiscal energy. Unlike the conventional approach of just relying on the trend or direction of price in order to extract profit, in vibratrading we also focus on extracting profit from price vibrations. The main advantage of this approach is that we do not need price to move in any particular direction in order to profit, as long as price continues to fluctuate.

The vibratrader derives profit in four ways:

1. Capital Appreciation
2. Oscillatory or Vibrational profits
3. Short Option Premiums
4. Stock/ETF Dividends.

You profit from capital appreciation if there is a rise in the price of your ETF or commodity. But in the process of waiting for price to rally, you may also derive consistent returns in the form of vibrational profits. By employing specific short option strategies, you can further add to your oscillation profits and transform the entire vibrational construct into a true non-directional trading methodology. Finally, you may also derive some passive returns in the form of dividends.

Nevertheless, the main purpose of vibratrading is to establish a basis of consistency in the extraction of returns from the market, regardless of the current value of invested capital. To reiterate, the primary goal is to continue to extract returns from the market indefinitely, with the aim of eventually reducing the cost basis of the stock or commodity to a risk-free investment. This implies that vibratrading is to be employed as an income strategy. Thinking of your trading as a business will help illustrate this important point.

VIBRATRADING AS AN INCOME STRATEGY

In vibration trading, your goal is to generate an income, preferably for life. You are aiming for the peace of mind and certainty of having a consistent income or "salary." Even though you have the potential to profit from capital appreciation, your objective is to generate regular cash flow. As such, capital gain is considered an added bonus.

As anyone running their own business can attest, the main objective of most ventures is to secure a steady and dependable return. A consistent income allows you to go about living life and planning daily activities. If you start a restaurant, your main objective is to make certain that you generate a monthly net profit, affording you a regular and steady income. You are not overly concerned about the fluctuations in the value of your assets as long as you manage to generate that monthly profit. In fact, there is nothing much you can do about the value of your assets; they are mostly subject to market forces. Nevertheless, you are still open to the prospect of selling off your restaurant should the value of your business assets or investments rise substantially, or if you receive an irresistible offer for your business. As mentioned, that would represent an added bonus.

> *You may never want to sell your restaurant, as you prefer the security, certainty, and "peace of mind" of a consistent income.*

In vibratrading, you invest in an ETF or commodity. Your "business" will be vibratrading that stock, and through fluctuations gaining a steady cash flow. As long as the stock or commodity exists, you need not be concerned with the overall value of your invested capital, since the maximum potential invested capital is known way in advance and is limited to a small and acceptable percentage of your total net worth. In this way, you generate an income that you may utilize immediately, allowing you to plan for your expenses, regardless of the current market value of the stock or commodity. As time passes, you will automatically reduce the cost basis of your total invested capital anyway. You need not wait for your investment to rise above your entry or breakeven price in order to generate an income.

In a sense, you have created an income mechanism, and your only concern is to make sure that it can carry on. Of course, if the street value of the mechanism itself rises, that is an added bonus. In any event, it will ultimately pay for itself and much more from the accrued income.

In vibratrading, we do not need to wait to recoup the value of our invested capital before spending the income it generates.

The obvious question is how could the investor or trader make sure that the stocks invested do not wind up? It does happen, though not very frequently. Nevertheless, this issue must be addressed in order to vibratrade effectively. The venues available for solving this are discussed extensively under the chapter on diversification. One of the methods discussed refers to the use of ETFs and commodities as our preferred trade and investment vehicles. This is because of ETFs' built-in stock replacement mechanism, which virtually ensures that their price will not fall to zero, unless every component stock winds up simultaneously. We shall employ numerous levels of diversification to address this issue. Through the components of vibratrading, it is easy to ensure that the price of the instrument traded not fall to zero, or *zero test level.*

INTRODUCTION TO THE COMPONENTS OF VIBRATRADING

Before getting into the implementation, let's start with an overview of the basic components and concepts that make up the vibrational constructs, scaling mechanisms, and the overall pyramidal structure.

In short, vibratrading is based on the structural interplay between three aspects:

- Price Averaging
- Scale Trading
- Trend Capture.

In price averaging, the trader or investor scales in, or buys, shares of a stock at various predetermined price levels which get successively lower as the price of the stock declines. These shares are later sold off for a profit once the price of the stock rises above the average price of all the shares purchased.

In conventional scale trading, the trader also scales in shares in the same way. The shares are then scaled out for a profit once the price rises to the next higher price level above the original buy price.

Some Basic Scaling Concepts and Definitions

Let us first familiarize ourselves with some preliminary scaling concepts.

Say we initiate our first long position at an entry price of $10. Let us also assume that we will enter longs positions at every successive $1 drop in price. That means that we should be buying into positions at $9, $8, $7 and down through $1, should price actually fall that far. We call $0 our zero test level.

In Figure 3.1, the length of the horizontal black line represents the position size for commodity traders, or *sharesize* for equity traders. A longer line represents a larger position or sharesize. So, if we buy the same amount of shares at each designated price level, the vibrational scaling construct should be represented by equal lines at equidistant $1 intervals from $10 to $0.

Now let's devise a scaling construct whereby we maintain the same entry price levels, but increase the position size of the stock at each lower entry price by 10 shares. It follows that our construct would now start to assume a bottom-heavy, pyramidal shape (see Figure 3.2).

Of course, if we devised a scaling construct with a diminishing position size at each successively lower entry price, the scaling construct would start to resemble a top-heavy, inverted pyramid instead.

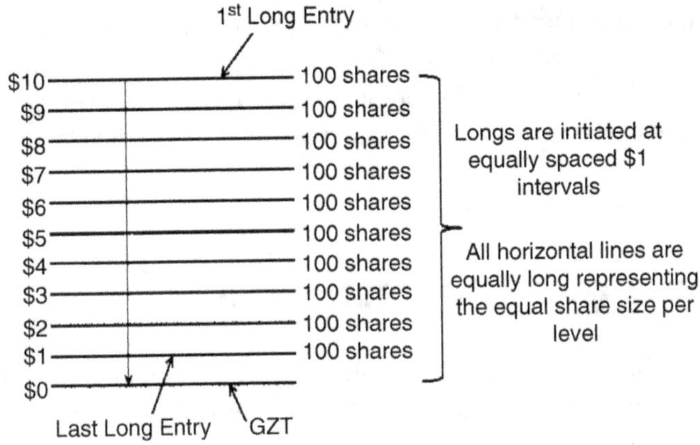

FIGURE 3.1 The Sharesize Diagram

The shape and size of the scaling construct is dependent on the:

1. Number of intervals
2. Distance between the intervals
3. Rate of increase or decrease in position size at each successive level, called the scale factor. (A scale factor of one means that all levels have the same position or sharesize. A scale factor of two means that each successively lower level's position or sharesize doubles.)
4. Effective range of the traded market action. (This may also be the historical range of the traded instrument.)

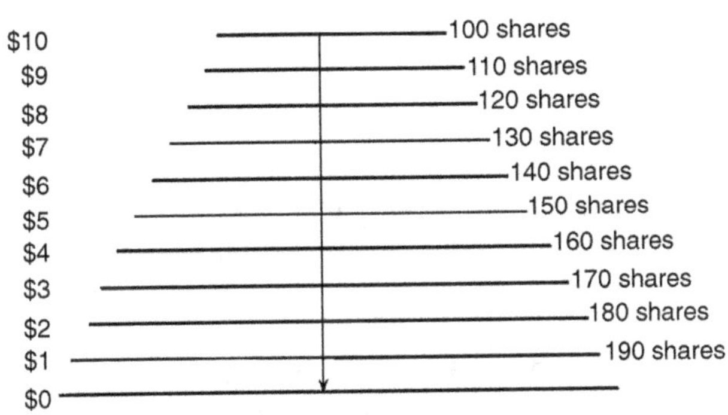

FIGURE 3.2 Bottom Heavy Sharesize Allocation

The manner in and amount with which we enter and exit the longs will determine the type of vibrational strategy employed and hence its profit structure. For example, scale factor and dual sizing affects the amount allocated to each level. This will be dealt with in more detail in later chapters.

MAIN COMPONENTS OF VIBRATRADING

Now that we have a basic understanding of scale trading, we can apply that knowledge to better understand vibrational trading. Vibratrading combines scale trading, price averaging, and trend capture to form:

1. Vibrational Structure
2. Vibrational Constructs
3. Vibrational Scaling Mechanisms.

1) Vibrational Structure

Bounded vibrational trading usually begins with an entry at the capstone entry level. However, bounded entries may be initiated anywhere within the pyramidal structure, from the pyramidal floor right up to the apex, which lies at the top end of the capstone.

The capstone entry level divides the vibrational pyramidal structure into two parts:

1. The Capstone
2. The Pyramidal Base.

Figure 3.3 is a graphical representation of the vibrational structure with the capstone situated above the pyramidal base. The capstone entry level lies between the two pyramidal structures. All bounded trades may be initiated anywhere between the pyramidal floor and the apex.

Significance of Entry within the Pyramidal Structure When we say that we are initiating a bounded entry within the pyramidal body, this means that we can afford to:

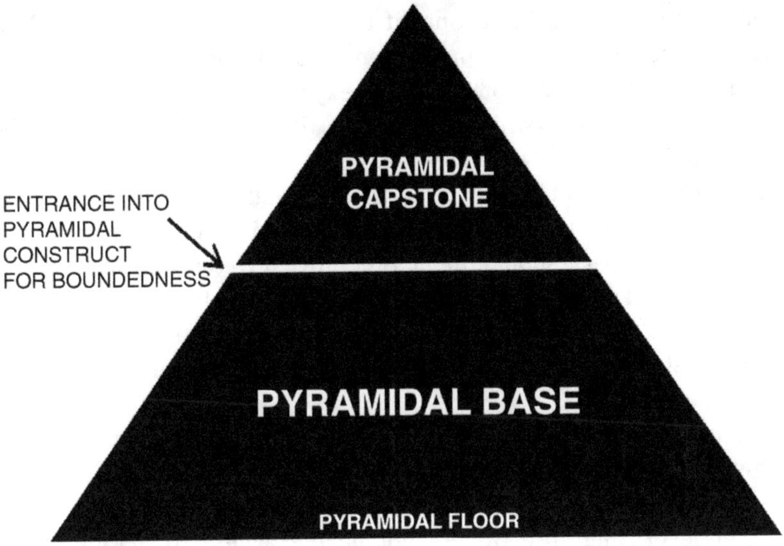

FIGURE 3.3 The Pyramidal Structure

1. Fund more long positions at levels below our initial entry, buying at each successively lower level down toward the pyramidal floor with sufficient disposable capital to hold all these positions until the zero test level, especially if trading on margin.
2. Fund more buying at levels above our initial entry, buying at each successively higher level, all the way up to the apex, with sufficient disposable capital to also hold these long positions until the zero test level.

The apex represents the highest price that we are able to afford within a chosen vibrational construct. It is situated at the top of our construct, at the point where all our capital would be fully allocated. Therefore, vibratraders would be unable to fund any more longs above the apex, but may hold on to any longs entered within the pyramidal body in order to capture some trend profits. A distinction must be made between entering positions at higher prices and entering positions at the apex. Although the bounded vibratrader is not able fund any new positions above the apex, it is still possible to initiate long positions at higher prices in a margin account if the previous entries are rendered breakeven and the new positions are entered via constant capital allocation per level. This creates a "floating apex" which allows the vibratrader to trade at higher prices while remaining at the apex. As we will see later, this is called the Free Swing method.

CFD traders would be able to fund more longs above the apex as they trade on margin. But, they would need to use a capital depleting mechanism like a stop loss to exit should the market move adversely against any long position initiated above the apex, as they would be unable to hold a position all the way to the zero test level. CFD traders will experience forced liquidation should price decline below the minimum margin percentage level.

Assuming that stock traders operate at full margin, they will not have the funds to enter a new long above the apex. The terms of capital boundedness require you to to maintain positions all the way to the zero test level, therefore traders not having access to margin trading will not be able to enter an unbounded long position. Traders on low margins may inadvertently initiate an unbounded long position. Not having enough capital to hold a long position until the zero test level could, in the worst case scenario, cause the forced liquidation of long positions at some point and an exit with a loss. For capital boundedness, no trades are allowed to exit with a loss. Hence, entering any position above the apex is unbounded.

The capstone entry level represents the price level from which we can afford to:

1. Fund further buys as price declines toward the pyramidal floor and simultaneously hold all long positions until the zero test level, especially if trading on margin.
2. Fund further buys up to the apex with sufficient funds to hold all long positions until the zero test level.
3. Scale out long positions up to the apex.

Put another way, the capstone entry level is any initial entry into the pyramidal structure whereby there will be sufficient capital to fund and hold long positions from the apex right down to the pyramidal floor.

Functionally, the pyramidal structure converts vibrational energy into fiscal energy. Note that in Figure 3.4, scaling mechanisms in the

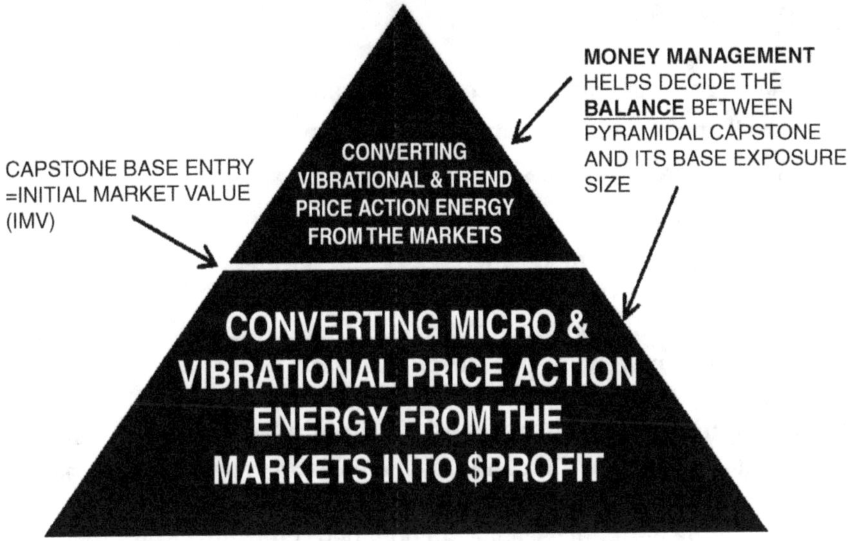

FIGURE 3.4 The Pyramidal Energy Structure

capstone also convert trend, or directional energy in the markets, into fiscal energy. The conversion of directional energy is also performed below the capstone entry level, if dual sizing (which will be explained later) is implemented.

The shape of the vibrational structure determines to a large extent the vibrational bias, and the speed and magnitude of vibrational returns. This will become clearer once we study and understand the specifics of vibrational constructs in later chapters.

2) Vibrational Constructs

All vibrational constructs are scaling or grid-type constructs and make up both the capstone and pyramidal base. Vibrational constructs that function above the capstone entry level consist of both capital bounded and order entry unbounded trend following and vibrational mechanisms. Constructs below the capstone entry level tend to be more of the vibrational type, though trend capture is possible with the implementation of the Macrosiso and dual sizing mechanisms.

The five main bounded vibrational constructs may comprise any or all of the following vibrational mechanisms: Microsiso, Macrosiso, Macrososi, Upside Macrosiso, and Upside Microsiso.

Their properties include:

1. Sharesize, scale factor, and capstone share multiples
2. Upsell and downbuy terminations
3. Interval size, intervals, and cumulative intervals
4. Working/total allocated capital and running cost
5. Specific construct modelling.

The actual definitions and treatment of the vibrational constructs will be covered extensively in subsequent chapters.

Concept of Boundedness and Trade Entry Type It is important to understand that all scaling mechanisms and constructs are based on the principles of boundedness, which will be discussed further throughout the book. In essence, all bounded trade entries are required to be long entries. Remember, there are no short entries in the bounded version of vibratrading.

3) Vibrational Scaling Mechanisms

Vibrational scaling mechanisms are the "building blocks" of the vibrational constructs which in turn make up the entire vibrational structure.

There are eight basic vibrational scaling operators:

1. **The Long Downbuy**: The scaling in of long positions as price declines.
2. **The Long Upsell**: The scaling out of long positions as price rises.
3. **The Long Upbuy:** The scaling in of long positions as price rises.
4. **The Long Downsell**: The scaling out of long positions as prise declines.
5. **The Short Downbuy**: The scaling out of short positions as price declines.
6. **The Short Upsell**: The scaling in of short positions as price rises.
7. **The Short Upbuy:** The scaling out of short positions as price rises.
8. **The Short Downsell**: The scaling in of short positions as prise declines.

It is important to note that in bounded vibratrading, all scale-ins must be long entries and all scale-outs long exits. The long and short upbuys and downsells are unbounded because they use stop orders.

MEANING OF THE SISO AND SOSI ACRONYMS

In this book, SI stands for "Scale In," meaning to enter a position. Similarly, SO stands for "Scale Out," meaning to exit a position. For bounded entries, SISO means that we buy low to open, and then sell higher to exit. SOSI means we first sell high to exit and later buy back lower to open. In SOSI, we enter at the capstone entry level with a long position and sell incrementally higher to exit.

Any long that exited across only one interval is called a Microsiso trade. The distance between each scale level is called the interval size (see Figure 3.5). Within this SISO mechanism, all exits are profit exits and only long entries are allowed to preserve boundedness.

The basic vibrational scaling mechanisms employed in vibratrading operate above and below the capstone entry level. These operators combine in various ways to form constructs, the basis of a true non-directional vibrational trading strategy, especially when combined with shorts option strategies.

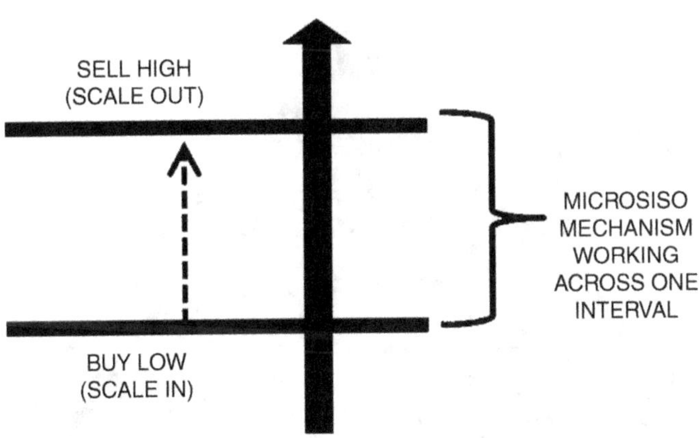

FIGURE 3.5 The Microsiso Mechanism

Constructs formed by these operators above the pyramidal capstone entry level include:

1. Bounded Upside Microsiso and Macrosiso
2. Unbounded Macrosimo (unbounded upbuy–upsell market order entry and exit)
3. Unbounded Upside Macrosiso
4. Bounded Macrososi.

Constructs formed by these operators below the pyramidal capstone entry level include:

1. Bounded Microsiso and Macrosiso
2. Bounded Upside Microsiso and Macrosiso
3. Unbounded Upside Macrosiso.

Notice that the Bounded Upside Microsiso/Macrosiso and Unbounded Upside Macrosiso are found both above and below the capstone entry level.

Figure 3.6 shows the various scaling mechanisms that reside both above and below the capstone entry level. The idea is to strike a trend to retracement balance in position allocation between the pyramidal capstone and its base entries.

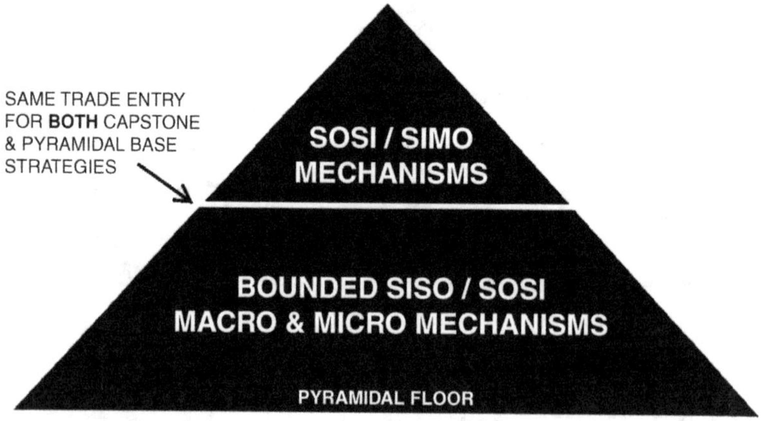

FIGURE 3.6 Locality of the SISO/SIMO Mechanisms

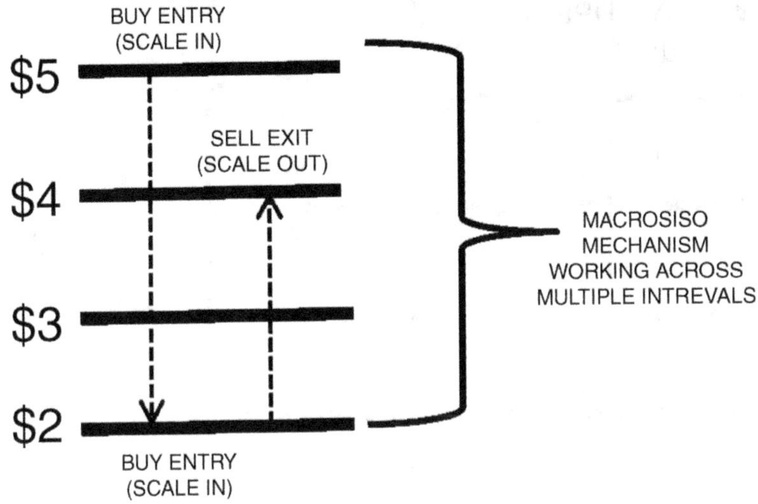

FIGURE 3.7 The Macrosiso Mechanism

BASIC SCALING ENTRIES AND EXITS

The rules of boundedness require that all downbuys be long and all upsells be profitable exits. Any long or short that exits across one interval is called a Micro trade. Any long or short that exits across two or more intervals is called a Macro trade.

Figure 3.7 shows a simple vibrational construct with four price levels making up the three scaling intervals. An interval is the distance between two adjacent price levels. As we can see from the illustration, we scale in new longs at specific levels as price falls. Each position is called a downbuy. We then exit, or scale out, as price rises above the corresponding scale level, and we call this an upsell. For example, imagine that we bought 100 shares at $5. We then buy, or scale in, another 100 shares at $4, then another 100 at $3, finally scaling in 100 shares at $2. We are only allowed to scale out or sell those shares at a price level above its original long entries. Therefore shares that we bought at $2.00 may be sold at the next higher level, at $3, and so on. We scaled in, and later scaled out at a higher price. Since we scaled in and then out across only one interval, we executed a Microsiso trade. If we had scaled in at $2 but scaled out at $4, we would have then executed a trade across two intervals, making it a Macrosiso trade.

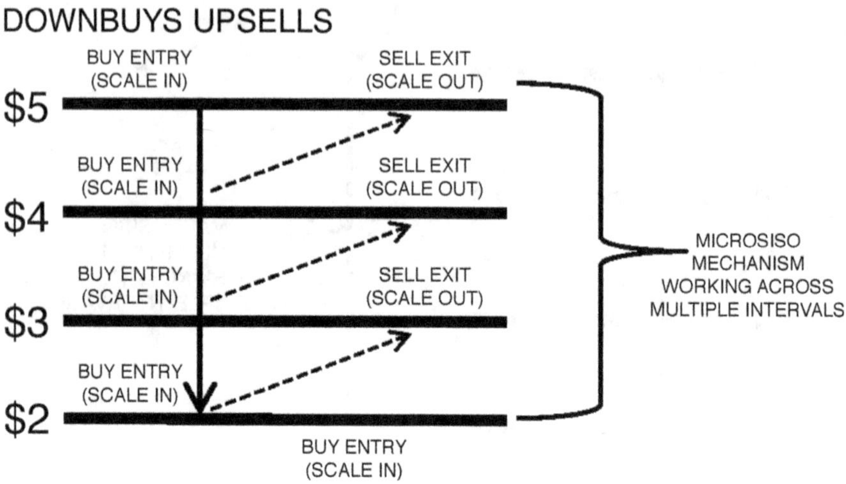

FIGURE 3.8 Microsiso Mechanism in Action

It is important to emphasize that in bounded trading we buy as price falls and sell as price rises. This is a strict requirement that derives directly from the principles of capital preservation. In other words, we only execute retracement trades: we exit via an upsell when price retraces or reverses up from a downbuy. The bounded vibrational constructs execute micro and macro retracement trades. In bounded vibratrading we never execute upbuys followed by downsells, which is unbounded trading.

In Figure 3.8, every upsell must be at a higher price level than its associated downbuy. In Microsiso, every upsell is executed one level above its original downbuy. You will learn in subsequent chapters about designing various vibrational constructs using these vibrational scaling mechanisms, which involve four main elements:

1. Total Capital Allocation
2. Scale Factoring
3. Intervals Sizing
4. Effective Price Ranging or Zoning.

The Upsell Rule

The size of the upsells must not exceed the total size of all the downbuys below it. If this rule is breached, then the vibratrader may end up with

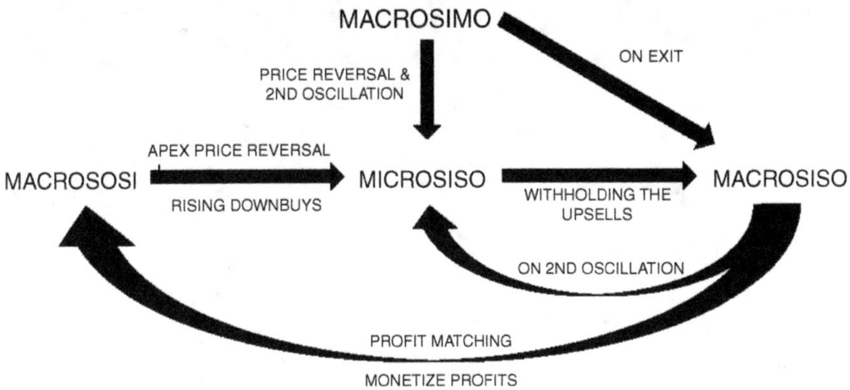

FIGURE 3.9 SISO-SOSI-SIMO Vibrational Transformation Pathway

an unintended loss or a net short position, which violates the rules of capital and directional boundedness and hence must be avoided at all costs.

The Interplay between Scaling Mechanisms

Vibratrading is extremely flexible and fluid in its application. Figure 3.9 shows a flow chart of the scaling mechanisms' transformative power.

CHAPTER 4

Controlling Risk in Vibratrading

In order for a trading methodology to be of any practical use, it must address the serious issues associated with risk in the markets. The true acid test of any long-lasting and robust trading methodology undoubtedly lies in its ability to effectively curtail the level of risk experienced by the portfolio, especially in drastic market downturns, including any Black Swan or Fat Tail events that may unfold rapidly.

This purpose of this chapter is to introduce the techniques and strategies used for controlling risk in vibratrading. Later chapters will provide a more detailed treatment on the construction and actual implementation of each of the specific risk control mechanisms mentioned in this chapter. We shall first list the types of risk that exist in the market, followed by the various control mechanisms that may be employed to manage those risks.

TYPES OF RISK

Generally, the word "risk" refers to several risk-related characteristics of a trade vehicle. High volatility leads to adverse price gapping and limit moves, while high leverage can lead to reckless over-exposure. Illiquidity is another risk, leading to serious trade disruptions. In the case of market shocks, low diversification can result in increased vulnerability. A high capital requirement, as its name implies, requires a larger capital exposure and thus, greater risk.

As far as vibratrading is concerned, there are five types of risk that may affect any portfolio or single stock:

- Risk of a broad market decline or collapse (Systematic Risk)
- Risk of an individual stock winding up (Specific Risk)
- Risk of an ETF or CFD broker winding up
- Risk of a large and permanent decline in the stock or CFD-based commodity price
- Risk of severe price volatility.

RISK CONTROL MECHANISMS

There are six main ways of managing risk in vibratrading, namely:

1. Trade instrument selection
2. Equity risk control techniques
3. Diversification-based strategies
4. Hedging strategies
5. The protective long put option strategy
6. Disposable capital.

I will illustrate, albeit qualitatively, how the risk control mechanisms work on each type of risk to protect the portfolio or individual stock.

Trade Instrument Selection

This is probably the most significant step that any investor can take to contain and manage risk in the portfolio. Even before attempting to apply any risk reduction techniques, the first thing any risk-conscious trader or investor should do is select a trading vehicle that is naturally predisposed to lower risk.

The vibratrader has a multitude of trading instruments to choose from nowadays, some being inherently more risky than others. Some possible choices are:

- Stocks
- Futures
- Mutual Funds
- Options
- Exchange Traded Funds (ETFs)
- Contracts for Difference (CFDs)
- Spot Foreign Exchange (FOREX).

The criteria by which a trade vehicle is chosen in vibratrading is directly related to its risk characteristics. This book will focus only on the safest tradable instruments: ETFs and commodity based CFDs.

The preferred vibratrading vehicle is an Equity ETF. It has excellent diversification characteristics, which any trader instinctively knows reduce risk. Additionally, popular ETFs are extremely liquid and not as capital intensive as futures trading—there is the possibility of participating in any ETF from as little as one share! This scalability and flexibility makes ETFs exceptionally accessible to the general public.

As you will see later, the vibratrader can continue extracting profits from the market and reducing cost basis indefinitely, regardless of whether the traded instrument's price returns to the original entry price level. Vibratraders, like scale traders, do not require the instrument's price to return to the original level, or even to rise above the average price in order to extract trend based profits. They only require that the traded instrument does not wind up, for if it does no trading will be possible.

The entire vibrational methodology is designed around the use of ETFs and CFDs. Of course, should the vibratrader insist on using conventional single stocks instead, there are some measures that try to replicate the functions of an ETF or commodity based CFD. However, this would be a more costly alternative. ETF replication will be covered in subsequent chapters.

Equity ETFs are chosen as one of the main trade vehicles in vibratrading due to two factors:

1. ETFs represent a large basket of component stocks, and as such, are very well diversified. The probability of every stock within an ETF winding up at the same time is negligible. As a result, it is the perfect complement to vibration trading, since the scale trading mechanisms allow trading to very low price levels, with the largest vibrational profits accrued at lowest price levels. It is therefore very important to choose a trade vehicle with a Net Asset Value (NAV) that has virtually no chance of going all the way south.
2. ETFs also have a built-in stock replacement mechanism that constantly replaces any stock that falls short of the ETF's minimum criteria for inclusion within its component basket of stocks. This will further ensure that the Equity ETF will maintain its Net Asset Value well above zero, again making it an ideal trading vehicle for scale trading the widest price range possible.

The unique ETF features described above are critical to the implementation of a vibrational trade. This is because in vibratrading, it is preferable that the trader has the opportunity to scale trade as wide a price range as possible, in order to reap greater profit from the lower price levels. And of course, the vibration trader wants to do this with the lowest probability of the trade vehicle winding up.

It is important to understand that even if an ETF actually crashed to zero, the trader would not suffer any catastrophic loss, except to the pre-agreed levels of risk allocated to that ETF. This is because it is a strict requirement in all scale trading methodologies to first assign, without fail, a predetermined maximum percentage of capital exposure in any one ETF. This is part of the equity risk control required in order to "scale trade" safely. Nevertheless, should an ETF be delisted, the vibratrader may horizontally reinvest the current market value of his or her investment in another ETF or commodity based CFD with a higher Price Leverage Ratio (PLR). This will help to maintain the effectiveness of the original investment.

Equity Risk Control Techniques

Equity Risk Control, or ERC, is a form of money management. The difference is that money management is only concerned with the

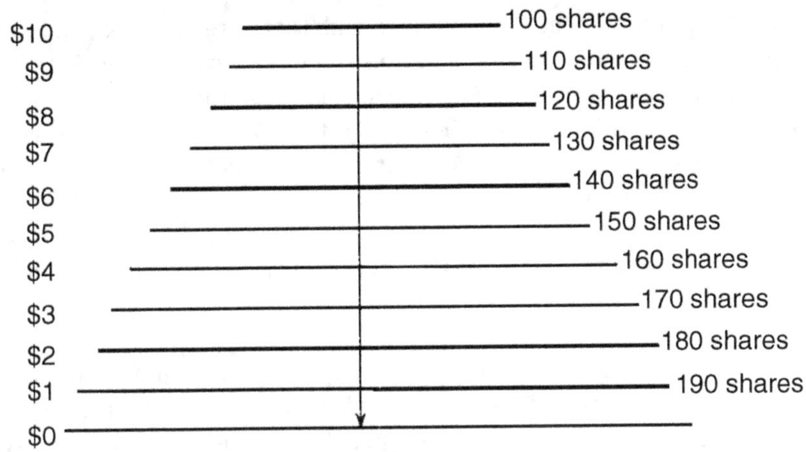

FIGURE 4.1 Vibrational Scaling Construct

percentage risk per trade, whereas ERC is concerned about the level of the overall equity in the account and how to contain the risks associated with a declining equity level. The vibration trader must decide on the maximum percentage of equity that is to be applied to the scaling construct, before selecting the type of scaling construct to be used.

An example will help clarify this.

Figure 4.1 depicts a very simple vibrational scaling construct with ten buy levels. The scale trader does not know exactly how many shares he or she will buy at each of those successively lower levels until a maximum allocation of capital is assigned to the selected ETF.

For the sake of illustration, let us assume that you have decided, in advance, to allocate a maximum of only five percent of your $143,000 trading capital to the selected ETF, which is now trading at $10. This amounts to a maximum capped exposure of 5 percent × $143,000 = $7,150. The scale trader then goes about configuring a very basic scaling construct, as illustrated in Figure 4.1, where the trader carefully pre-plans to buy a specific the number of shares at each successively predetermined lower price level, for a maximum capped investment of $7,150. Hence if the ETF actually winds up, the trader will only lose five percent of his total available equity. As long as the ETF does not wind up, the trader will be able to consistently extract vibrational profits from the ETF's price action across the entire trading range. Even if price fails to rise above the entry or average price, scale traders can still extract profits until they have reduced the potential cost basis of the entire investment. Therefore, the risk of a large and permanent decline in price is of no real consequence to the vibration trader. In fact, the vibration trader expects a fall in price in order to scale trade the deeper price levels, where the returns are the highest. This is because of increasingly higher share purchases at successively lower price levels, which allows the largest oscillation profits. Options strategies will also be incorporated throughout the scale construct to further supercharge the returns and afford the trader some downside returns to reduce drawdowns.

Risk of Multiple ETF Failures To expand on this, if vibration traders decide to invest in four separate ETFs, they will repeat the process, allocating the maximum allowable risk per ETF. Therefore, if a trader now decides on a maximum two and a half percent exposure per ETF, this will translate into a potential loss of only ten percent of the total available equity. But, again, this loss can only occur if every single component stock in all four ETFs winds up at the same time.

So, it is vital in vibratrading, or any form of scale trading, that the total capital exposure for any investment be capped at a predetermined maximum value from the very beginning. This will ensure that in the event of an ETF winding, the vibratrader will know, in advance, the maximum risk associated with that ETF, thereby removing all uncertainty.

The entire process is meticulous and precise. There is absolutely no room for a frenzied bottom fishing expedition in vibratrading!

Risk of Price Volatility In standard directional trading, losses accrue very rapidly during periods of extreme price volatility. This loss is directly attributed to the triggering of stop losses. Hence risk increases with price volatility.

In vibratrading, on the other hand, the more volatile or violent the price action is, the more oscillation profits are made. In fact, price gapping and limit moves potentially generate more profits for the bounded vibrational trader. Risk actually decreases with price volatility, because, in bounded scale or vibration trading, price gapping can only result in two possible outcomes. Gapping beyond the planned exit price level provides the longs positions with additional profit. Gapping beyond the buy limit entry level gives the vibration trader an opportunity to buy at a lower price, thereby increasing his potential profit once price begins to rise.

Aside from an ETF hitting ZTL, almost every other scenario presents the vibratrader with an opportunity to profit. That is one of the greatest advantages of bounded vibratrading.

An ETF's Cost Basis Even if an ETF does hit the ZTL, there is no guarantee that the trader will suffer a loss of capital, as the trader may only be losing the "risk-free" oscillation profits. For example, if the trader has already reduced the cost basis of the maximum potential allocated capital or investment, that vibratrader is now extracting additional, risk-free oscillation profits. Once the accrued oscillation profits exceed the maximum potential allocated value, the entire investment is said to be totally risk free, or at break even. Once risk-free status is achieved, a decline to the ZTL can result only in a loss of profit, not the original seed capital.

ETF Delisting When an ETF is delisted, the trader or investor is returned the equivalent cash value of all the shares held in that ETF, along with any outstanding dividends. There may also be an option to purchase some of the actual stocks previously held if there is a large number of shares redeemed. Therefore, delisting does not adversely affect the vibratrader, as it presents the option to horizontally reinvest the funds (and carry on trading vibrationally) in similar ETFs, preferably with higher PLRs.

All things being equal, the risk of getting stopped out in directional trading is phenomenally higher than that of an ETF winding up. Traders of individual stocks do not enjoy the same level of protection that an ETF can provide.

Diversification-based Strategies

Even though we have already discussed the use of ETFs as the Vibratrader's preferred trade vehicle, this only represents the first level of diversification. We took advantage of the ETF's diversification properties and the inbuilt stock replacement mechanism to help prevent the traded price from experiencing a Zero Test Level event. That is the most important requirement for vibration or scale trading.

There are, however, five other levels of diversification that a vibratrader can use to control risk.

The Second Level of Diversification: ETF Size & Type At the first level, we diversified away our risk by investing in a replenishing basket of stocks. At the second level, we will be selecting the size and type of ETF diversification to further reduce the possibility of an ETF hitting zero test level.

For added protection, it is better to select an ETF with a larger basket of component stocks. This will spread the risk more evenly, reducing volatility as well as the possibility of winding up. Relatively, it is more difficult for an ETF with a basket of 500 stocks to wind up simultaneously, compared to an ETF with a 10-stock basket.

Scale trading in commodity futures has been extremely popular for many years. But with the advent of diversified funds, such as ETFs and higher leveraged commodity-based CFDs, where the minimum purchase is just one share, scale traders can finally migrate to a more affordable platform, avoiding the large capital and margin requirements associated with conventional commodity futures trading.

The reason commodity futures became so popular with scale traders is that commodity prices are not expected to fall to zero, as there is a constant demand for those commodities. This creates a natural support, or price floor, from which the scale trader may plan his scale trading range. Therefore, knowing that commodity prices will never fall to zero, we can further protect our investment if we incorporate commodity based ETFs or CFDs into our portfolio.

Before discussing the next level of diversification, I should point out that with ETFs, we can effectively extend that price floor right down to the ZTL. This is possible because of four important ETF qualities:

1. Affordability (resulting in very low maximum allowable risk allocations per ETF).
2. Easy participation (shares are scalable, starting from as low as one share).
3. Excellent diversification characteristics (large baskets of stocks and markets available).
4. Built-in self-replacing stock mechanism for added protection (further reducing possibility of a ZTL event).

The Third Level of Diversification: ETF Quantity Initially, we selected the ETF for its diversification features. At the second level of diversification, our preferences leaned more toward commodity based ETFs, as well as larger ETFs, for added protection. We now begin to diversify on an even higher level, that is, by investing in more than one ETF.

Multiple ETFs elevate the level of protection against a zero test level event even further. At this level of diversification, the trader must allocate, in advance, the maximum allowable exposure for the entire ETF portfolio. So, in the extremely unlikely event that every ETF in the portfolio winds up, the trader's total potential loss is capped at acceptable levels. The entire portfolio may, for example, only reflect 20 percent of the trader's entire net worth.

The Fourth Level of Diversification: Inversely Correlated ETFs The prudent vibration or scale trader should construct the optimum combination of ETFs within the portfolio to further diversify away the risks involved. This is done by selecting a combination of ETFs that are inversely correlated. Inversely correlated means that when a drastic downturn is experienced in one market, sector, or industry, another area may experience a rally or upturn. I illustrate inverse correlation with examples in the chapter on diversification. This will help to curtail the systematic risks associated with a global or broad market collapse.

It must be noted that these inverse correlations can and often do break down, especially in the face of extreme fear or irrationality in the markets. Markets that tend to move together start to "decouple," while some that usually move in opposite directions start to synchronize and track each other. It is generally accepted nowadays that most markets tend to be correlated during a global shock event or crisis. To help reduce these correlation breakdown effects, we must explore the ways in which these correlations arise. Markets are inversely correlated in two ways: Market Driven Correlation and Structurally Driven Correlation.

Market driven correlation is governed by market forces. Market forces are, to a very large extent, determined by human perception. Hence natural correlation, be it direct or inverse, is subject to human emotion, which is clearly evidenced during a sudden or insistent market downturn. As such, these types of correlations are vulnerable to a breakdown in normal intermarket relationships.

On the other hand, structurally driven correlations are mathematically based. They are not as susceptible to large intermarket breakdowns or decoupling. Reducing the correlation breakdown effect will be dealt with in later chapters.

The Fifth Level of Diversification: ETF Oscillatory Propagation At this level, we introduce the concept of Oscillatory Propagation. The actual calculations for setting up the strategy will be covered in detail in the chapter on diversification.

In oscillatory propagation, we set up multiple ETFs so as to use price fluctuations to help create capital for the purchase of additional ETFs. This new capital is created via an inter-level price oscillation within the scaling construct, and this single oscillation is sufficient to fund the purchase of another ETF stock. In this way, vibratraders create new investments for their portfolio, courtesy of market action. The speed of propagation of these new ETFs is a function of price oscillation, or market volatility.

Therefore, the more volatile a market is, the faster it will propagate the risk-free purchase of these new ETFs, which are totally funded by oscillation profits. Using this system has two immediate benefits:

- Totally removes the cost of purchasing new ETFs
- Creates a larger set of ETF scaling mechanisms which will generate larger profits.

The Sixth Level of Diversification: ETF Currency Risk At the sixth level, we understand that no matter how much profit is made from vibration trading, the value of our current total equity is still subject to currency risk. No amount of inversely correlated instruments will be able to curtail the loss should the value of the base currency collapse. One solution would be to trade commodity based ETFs or CFDs. If the value of the United States Dollar (USD) depreciates, this will inflate the value of commodities, thereby helping to reduce some of the adverse effects of the USD decline in purchasing power.

Therefore, savvy traders may currency hedge their investments, in order to best preserve the long-term value of equity in the trading account.

Hedging Strategies

Since the vibration or scale trader has already committed to a maximum capital allocation per ETF, incorporating a simple hedging strategy will generate additional returns.

In the following chapters, the reader will be shown various hedging strategies whereby additional returns are generated via the application of shorts in the market. But the ratio of longs to shorts must be designed such that the overall position of the ETF is always net long. From that point, any short profits derived will only add to the scale trader's original downside profits. Of course, some upside trend profits will be sacrificed in the process. Even in the event of a "shorts recall," which occasionally occurs, the vibration trader is not disadvantaged in any way. This is because the short profits represent only additional profit.

Protective Long Put Option Strategy

The trader may also incorporate a protective long put option strategy for the purposes of both reducing the losses associated with a full-blown ETF meltdown and generating additional profit during the vibration trading process. How far out of the money the long puts are depends on the amount of risk-free fund available. The more funds available, the less the long puts will be out of the money.

The long put option should be funded via oscillation profits. It may also be funded via short profits, derived from the hedging strategy mentioned earlier. That is one of the main purposes for initiating the hedging strategy. But remember, buying long options is an unbounded activity if it is not funded by vibrational profits. This is because if price does not move favorably, the vibratrader will suffer a loss of capital from losing the option's premium. Otherwise, establishing a long put is an effective technique for managing equity drawdowns.

The Disposable Capital Rule

Finally, vibratraders are advised to use disposable capital for their investments. This is to ensure that traders and investors only work with funds that they can afford to allocate to the methodology indefinitely. Once the downbuys are initiated in a vibrational construct, there may be a period of drawdown. Any exit during this drawdown period will lock in a loss. The entire amount may be horizontally reinvested to restart the vibratrading process and circumvent this earlier "loss."

We have just introduced the various methods and strategies used in vibratrading for the curtailment of risk. You should have at least a cursory understanding of most situations where risk arises, including risks associated with:

- broad market decline or collapse
- individual stock winding up
- ETF winding up
- large and permanent decline in ETF/commodity based CFD prices
- severe price volatility.

A more detailed discussion on each and every technique mentioned in this chapter will be provided as you work through the book and begin enacting the lessons learned.

CHAPTER 5

The Mechanics of Equity-Based Price Action

This chapter introduces the trader to the nuts and bolts of price and ratio valuation which are important considerations in the design of vibrational constructs. It covers simple profit and loss, market value, and leverage ratio calculations which form the basis for selecting the most appropriate ETF or CFD.

$$\begin{aligned}
\text{Current Market Value (CMV)} &= \text{Total Shares} \times \text{Current Price} \\
\text{Initial Market Value (IMV)} &= \text{Total Shares} \times \text{Entry Price} \\
\text{Profit/Loss(P/L)} &= \text{CMV} - \text{IMV} = \text{Total Shares} \\
&\quad \times (\text{Current Price} - \text{Entry Price})
\end{aligned}$$

EQUITY-BASED CALCULATIONS

Before we can grasp the finer points of vibrational trading, we need to be somewhat familiar with equity-based profit and loss calculations. Let's take a look at how to calculate the amount of gain or loss for every $1 move for a given number of shares.

Example 1

Assume that we are long 100 shares (one lot).

So, P/L for a \$10 to \$11 change in price is CMV − IMV:

$$\$1,100 - \$1,000 = \$100$$

Therefore, the value of a dollar move is \$100.

This happens to be the number of shares we are currently long (see Figure 5.1).

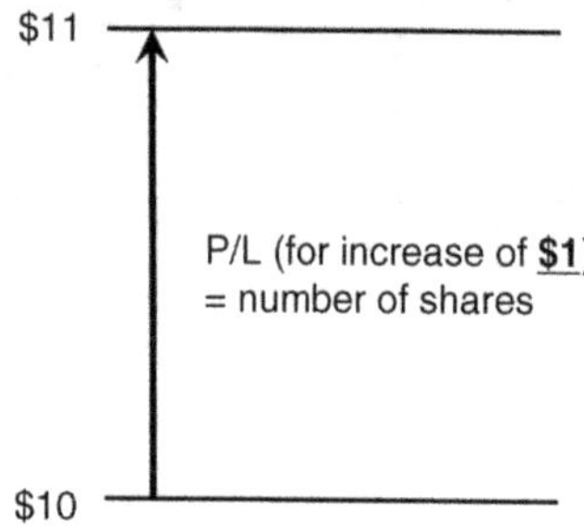

FIGURE 5.1 Measuring P/L

Example 2

Let us now assume that we are long 35 shares.

The P/L for an \$11 to \$10 drop in price is

$$\$350 - \$385 = -\$35$$

Therefore, the value of a dollar move is −\$35.

This again happens to be the number of shares we are currently long.

Therefore:

$$\text{Value per \$1 Move} = \text{Share Size}$$

We see that the running capital or P/L for every \$1 move in price is simply equal to the number of shares held.

MARKET VALUE VS. PROFIT POTENTIAL

Two traders invest \$100 in the markets, the only difference being they bought into the markets at different prices. Trader A bought 25 shares at \$4,

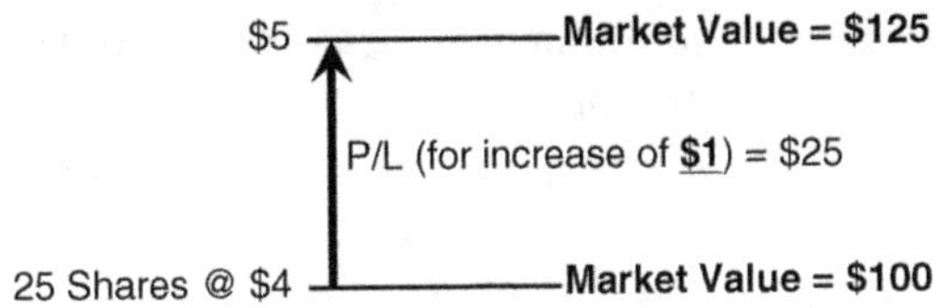

FIGURE 5.2 High Profit Potential

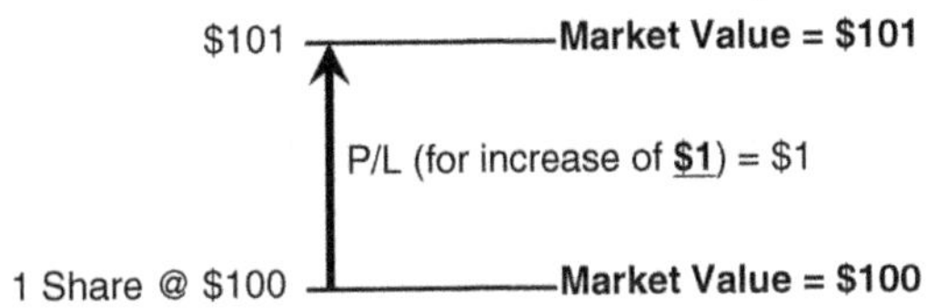

FIGURE 5.3 Low Profit Potential

and Trader B bought 100 shares at $1. They both invested an IMV of exactly $100 (refer to Figures 5.2 and 5.3.)

Trade A buys 25 shares for $100.

Trader B buys 100 shares for $100.

The significant difference between these two traders is the profit potential. Trader A has a higher profit potential: a dollar increase in price results in more gain. Trader B has the potential to profit or lose $100 per every $1 move in the markets, whereas Trader A will only be exposed to a quarter of that value, even though both traders invested the same IMV of $100. Therefore, it is not the amount that you invested in the stock, but the number of shares that makes all the difference.

Market value does not tell you the profit potential of your positions!

PRICE LEVERAGE RATIO (PLR)

Let the entry price be EP, current price be CP, and market value be MV. Using those values, we can calculate the percentage P/L based on a change in market value from IMV to CMV.

$$\begin{aligned}\text{Percentage Change in P/L} &= (\text{CMV} - \text{IMV})/\text{IMV} \times 100\% \\ &= \text{Shares} \times (\text{CP} - \text{EP})/(\text{Shares} \times \text{EP}) \times 100\% \\ &= (\text{CP} - \text{EP})/\text{EP} \times 100\%\end{aligned}$$

We see that the change in our P/L for any particular trade is simply the change in price over the entry price. This base formula gives us a ratio, and multiplying by 100 percent changes the solution into the percentage P/L.

We see that as entry price declines, the changes in P/L will start to increase proportionally. This is called the price leverage ratio. It is defined as the absolute change in PLR (that is, it is always a positive value).

$$\text{PLR} = \text{ABS Change in P/L} = \left|\frac{\text{CP} - \text{EP}}{\text{EP}}\right| \qquad \text{(in Ratio Form)}$$

Example 1

Let CP = \$101 and EP = \$100.

PLR = ratio change in P/L (per \$1 move) = (101 – 100)/100 = 0.01.

Example 2

CP = \$2 and EP = \$1.

PLR = (2 – 1)/1 = 1.

For an equal dollar move, the ratio change in P/L increases as EP decreases.

Let us now examine how the PLR changes with an equal change in price at progressively lower EP values.

Example 3

CP = \$20 and EP = \$10.

Ratio change in P/L (per \$10 move) = (20 – 10)/10 = 1.

Example 4

$$CP = \$15 \text{ and } EP = \$5.$$

$$\text{Ratio change in P/L (per \$10 move)} = (15 - 5)/5 = 2.$$

Example 5

$$CP = \$11 \text{ and } EP = \$1.$$

$$\text{Ratio change in P/L (per \$10 move)} = (11 - 1)/1 = 10.$$

We are exposed to the greatest P/L changes at lower entry prices. For a single dollar fall in price, our earning potential increases. This price leverage effect is a natural mathematical edge that must be exploited!

The Unit Price Leverage Ratio (UPLR)

Unit Price Leverage Ratio, or UPLR, is defined as the ratio change in P/L for a fixed \$1 move in EP.

$$\text{Unit Price Leverage Ratio (UPLR)} = \$1/EP$$

To explain this ratio, I have calculated the UPLR for various values of EP.

Example 1

$$\text{If } EP = \$1, \text{ then } UPLR = \$1/\$1 = 1.$$

Example 2

$$\text{If } EP = \$2, \text{ then } UPLR = \$1/\$2 = 0.5.$$

Example 3

$$\text{If EP} = \$50, \text{then UPLR} = \$1/\$50 = 0.02.$$

Example 4

$$\text{If EP} = \$100, \text{then UPLR} = \$1/\$100 = 0.01.$$

Relationship of PLR to UPLR

In the equation below, we see that another way of defining PLR is to multiply the UPLR by the absolute change in price.

$$\begin{aligned} \text{PLR} &= \left|\frac{\text{CP} - \text{EP}}{\text{EP}}\right| \\ &= \frac{1}{\text{EP}} \times |\text{CP} - \text{EP}| \\ &= \text{UPLR} \times |\text{CP} - \text{EP}| \end{aligned}$$

Example 5

$$\text{If price rises from } \$10 \text{ to } \$15, \text{PLR} = (15 - 10)/10 = 0.5.$$

$$\text{Similarly, UPLR} \times (\text{CP} - \text{EP}) = (1/10) \times \$5 = 0.5.$$

Example 6

$$\text{If price falls from } \$10 \text{ to } \$3, \text{PLR} = (3 - 10)/10 = 0.7.$$

$$\text{Similarly, UPLR} \times (\text{CP} - \text{EP}) = (1/10) \times \$7 = 0.7.$$

In order to increase our profitability, we should position ourselves at the lowest EP available for the greatest returns. That way, we only need a small price move to experience a large percentage return.

MONEY LEVERAGE RATIO (MLR)

Money Leverage Ratio (MLR) is the potential leveraging effect of PLR on the ratio of capital invested. It is used in determining your final dollar P/L.

$$\text{MLR} = (\text{Ratio Change in P/L}) \times \text{Ration Capital Invested}$$
$$\$\,\text{P/L} = \text{MLR} \times \text{Total Equity}$$

EXAMPLE 7

Let us assume that CP = \$15, EP = \$10, and IMV = 5% of equity where the total equity is \$100,000.

$$\begin{aligned} \text{MLR} &= \text{PLR} \times \text{Ratio of Capital Invested} \\ &= (\text{CP} - \text{EP})/\text{EP} \times (\text{IMV}/\text{Total equity}) \\ &= (15 - 10)/10 \times 0.05 = 0.5 \times 0.05 = 0.025 \end{aligned}$$

Hence, the final dollar profit and loss is 0.025 × \$100,000, or \$2,500.

More capital invested at lower prices increases your potential profit; put another way, lower EP coupled with higher IMV produces greater P/L. This means that your MLR is higher with a lower PLR, even without investing more capital in a trade. Nevertheless, the more you invest per trade at lower prices, the higher your potential returns.

Remember that higher IMV at lower EP increases your profit drivers as well, an important point which underlies the theory behind the vibrational constructs. At low prices, both buying leverage (our next topic of discussion) and PLR will amplify returns. In other words, you have more profit drivers, or shares, positioned at lower prices.

BUYING LEVERAGE RATIO (BLR)

BLR represents the capacity to buy more shares at lower prices for an equal amount of capital or IMV invested.

Using the formula below, we can determine that as CP decreases, the buying power increases. We already know that as EP decreases, the PLR starts to increase. Therefore, the ideal is to seek out low EP values with declining prices. This increases the BLR and PLR simultaneously.

$$BLR = \frac{EP}{CP}$$

The ratio of shares buyable is equivalent to the BLR, the ratio of EP to CP. In the below example, the number of shares buyable at $5 is 2.4 times the original amount of shares buyable at $12.

Example 8

Assume invested capital is $12,000.

The number of shares buyable at $12 = $12,000/$12 = 1,000 shares.

Number of shares buyable at $5 = $12,000/$5 = 2,400 shares.

Therefore, the Ratio of Shares Buyable = (2,400/1,000) = 2.4, meaning our buying power has increased by 2.4 times. To check our BLR equation, BLR = $12/$5 = 2.4.

In Tables 5.1 and 5.2 I've illustrated a comparison of two price activity ranges, the first with a price range of $99 and the other $1.98. We see that most of the ratios and quantities are identical, except for the pronounced difference in UPLR values.

TABLE 5.1 BLR and Comparison for an activity range of $99

	PRICE	PRICE DECLINE (PLR)	SHARES BUYABLE	BLR	UPLR
↑	EP = $100	0	100	1	0.01
$99 RANGE	CP = $50	0.5	200	2	0.02
	CP = $20	0.8	500	5	0.05
↓	CP = $1	0.99	10000	100	1

TABLE FOR FIXED IMV of $10,000

TABLE 5.2 BLR and PLR Comparison for an Activity Range of $1.98

	PRICE	PRICE DECLINE (PLR)	SHARES BUYABLE	BLR	UPLR
↑	EP = $2	0	100	1	0.5
$1.98	CP = $1	0.5	200	2	1
RANGE	CP = $0.40	0.8	500	5	2.5
↓	CP = $0.02	0.99	10000	100	50

TABLE FOR FIXED IMV = $200

By studying the bottom row in Table 5.2, we can see that at $0.02, we can buy 10,000 shares for a $200 investment. This represents 100 times more shares buyable at this lower price compared to EP, so BLR is 100.

If price goes up to $1.02, we profit = $1 × 10,000 shares or $10,000. Our PLR should confirm this. Since UPLR at $0.02 is 50, for every dollar increase we will make 50 times the investment. So, if we invest the same $200 at $0.02, our profit after a move up to $1.02 is 50 times $200, or $10,000.

Example 9

Let's assume that you bought 10,000 shares of a stock. How many more shares could you have bought if price declined by 20 percent for the same investment?

$$\text{BLR} = \text{EP/CP} = 1/0.8 = 1.25.$$

$$\text{Shares buyable} = \text{BLR} \times \text{Original amount}$$
$$= 1.25 \times 10000 = 12500 \text{ shares}$$

Therefore, you could have bought 2,500 more shares than the 10,000 purchased for an increase of 25 percent.

The Non-Linearity of BLR

BLR does not behave linearly across prices. For example, a $9 change in price from a lower PLR compared to the same $9 change at a higher PLR is vastly different:

PLR at $10 = ($10 − $1)/$1 = 9.

On the other hand, PLR at $100 = ($100 − $91)/$91 = 0.1

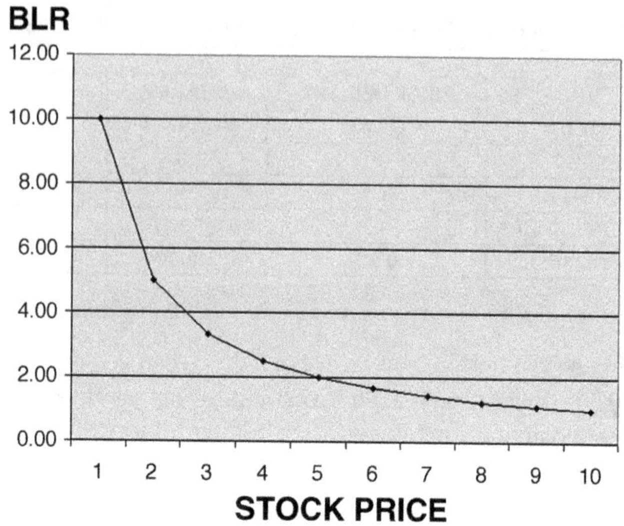

FIGURE 5.4 High BLR at Low Share Prices

For the same \$9 decline in price, (BLR = \$10/\$1) 10 times more shares can be bought at \$1.

The BLR becomes 1.1 (100/91) in the second example, meaning only 1.1 times more shares can be bought at \$91. So, 10/1.1 equals 9.1 times more buying power in the lower PLR, which proves that buying power increases exponentially at progressively lower prices with respect to the same price change.

To further illustrate this point, Figure 5.4 displays stock prices rising from \$1 to \$10, with the BLR at a value of 10, or 900 percent, in the first dollar. The second chart, Figure 5.5, displays an equal rise in stock price, but from \$91 to \$100. Here we see that the BLR was only 10 percent higher before the same \$9 rise in price.

It is not the amount of capital invested that determines how much you make, but the number of shares you have. This makes BLR critical in vibrational trading. The more profit drivers you can buy, the greater your profit potential during a price move. Lowering prices increase your buying power, and therefore your potential returns. For example, spending \$1,000 on a single share has a much lower profit potential than spending the same \$1,000 on 1,000 shares. A price move of \$1 will give you only a \$1 profit, compared to \$1,000 profit for shares that you bought at \$1.

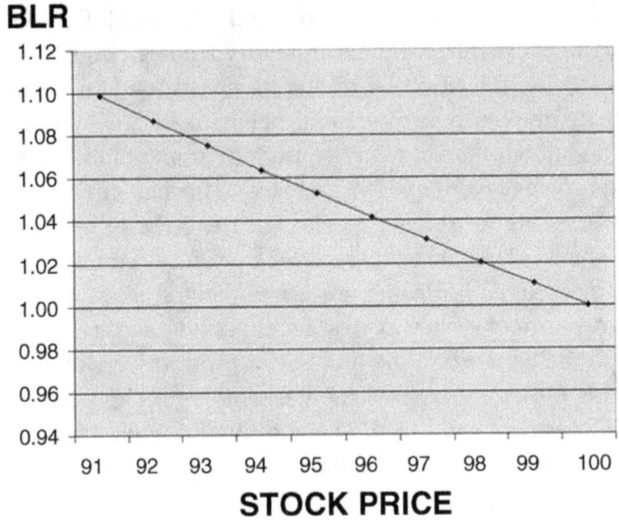

FIGURE 5.5 Low BLR at High Share Prices

Significance of PLR and BLR

In bounded vibrational trading, we extract profit via an oscillatory scaling mechanism consisting of downbuys and upsells. The more profit drivers we possess, the greater the vibrational returns. Incidentally, there is no oscillational loss in bounded vibratrading.

PLR tells us we need to seek out low entry prices to experience greater percentage returns. BLR also indicates that we should seek out lower entry prices so that our buying power is greater, which in turn influences the number of profit drivers we are able to invest in. MLR proves that our capital need not work so hard if both the PLR and BLR are high. On the other hand, it also tells us that the more we invest, the greater our potential returns.

At this point it must be noted that BLR is of little advantage in commodity futures trading, as there is already an overall increase in buying power by virtue of the high leverage offered. In addition, these traders do not benefit from the PLR as returns are based on relatively fixed margins. Futures traders do not buy more at lower prices because there is no advantage in doing so, save taking advantage of some potential demand at historic lows. On the bright side, futures traders are not forced to risk more capital at higher prices. The variation of PLR and BLR is fairly linear throughout the price range, unlike in equity trading, where the trader is required to pay the full price for taking up a position and therefore these factors display non-linear, exponential behavior.

One little-known reason many equity day traders fail is low BLR and PLR values, especially at higher share prices. An equity day trader has less buying power at higher prices, which will result in a significantly lower number of shares. The equity day trader therefore requires greater price excursions to make up for losses sustained while trading with more profit drivers at lower prices. One way to solve this asymmetric leveraging effect is to proportionally increase the number of profit drivers at higher prices. However, this directly increases the percentage of capital risked, unless tighter stop losses are employed. The commodity futures trader is happily spared, to a large degree, from this asymmetry.

ACCOUNT LEVERAGE RATIO (ALR)

Account Leverage Ratio, or ALR, affords the trader the opportunity to trade without paying the full market value of the position. Instead, only a deposit is required, at a fixed percentage of the original market value traded. This deposit is called the initial margin, or \$Margin. The maintenance margin is the amount of money required to hold the position indefinitely. A margin call will be issued should your available trading funds fall below the maintenance margin level. You would then need to replenish the account's equity to the minimum level required by the initial margin within a given period to avoid liquidation.

$$\text{Account Leverage Ratio} = \frac{\text{Full MV}}{\text{\$Margin}} = \frac{100}{\text{\%Margin}}$$

Percentage Margin Level

Before we delve into percentage margin level calculations, we need to define a few commonly used terms.

The account balance is the amount of capital in the trading account, not including all open positions' P/L. Used margin is the total amount of margin used in the trading account. Unused margin is the amount of capital left in

Example 10

Let us assume that an initial margin of $2,000 is required to trade one standard lot of spot USDJPY (US Dollars/Japanese Yen), which has a nominal value of USD100,000.

$$\text{ALR} = \text{CMV}/\$\text{Margin} = \$100,000/\$2000 = 50.$$

$$\textbf{Margin Percentage} = 100\%/50 = 2\%.$$

The spot FOREX trader only needs to deposit 2 percent, which is 50 times less than the nominal amount required to trade a standard lot. This percentage is also called the margin percentage. Without account leverage, the spot trader would need to place $100,000 in order to trade a single standard lot. It is especially important to note that initial margins may be adjusted by the broker to absorb market volatility or illiquidity, very often without prior notice. This is not a problem since we are not forced to depend on margin in bounded vibratrading. I cover that and further principles of boundedness in the next chapter.

the account, reflecting all open positions P/L, less the total used margins. Equity represents the true value of the entire trading account, inclusive of the balance, all open positions' P/L, and used margin.

$$\begin{aligned} \text{Useable Margin} &= \text{Equity} - \$\text{Margin} \\ \%\text{Margin Level} &= (\text{Equity}/\$\text{Margin}) \times 100\% \\ &= (\text{Balance} + \text{Open P/L})/\$\text{Margin} \times 100\% \end{aligned}$$

Please note that percentage margin level is distinct from margin percentage. Margin percentage is the amount needed to enter a trade, based on the percentage of the full traded market value. Percentage margin level, on the other hand, is the percentage level of equity to used margin. Many online brokers will issue a margin call (an urgent request for capital) at or under the 100 percent level, where long positions will begin liquidation. However, some close open positions without issuing any margin calls once the percentage margin falls below the cutoff level. As part of capital boundedness, we should be aware of our percentage margin level at all times.

EXAMPLE 11

What is the total percentage margin level of the account if a trader intends to open two trades with differing margins?

Equity = \$20,000, \$Margin 1 = \$2,000, \$Margin 2 = \$6,000

$$\begin{aligned}\%\text{Margin Level} &= \text{Equity}/\text{Total \$Margin} \times 100\% \\ &= \$20,000/\$8,000 \times 100\% \\ &= 250\%.\end{aligned}$$

EXAMPLE 12

%Margin Level has been set by a certain brokerage at 100 percent. Can you still trade?

\$Margin = \$3,000, P/L = −\$8,000, Balance = \$10,000

$$\begin{aligned}\%\text{Margin Level} &= (\text{Balance} + \text{P/L})/\$\text{Margin} \times 100\% \\ &= \$2,000/\$3,000 \times 100\% \\ &= \text{about } 67\%.\end{aligned}$$

Since your current %Margin Level is below the minimum, you are not permitted to initiate any further trades.

In the following example, we will briefly ignore the %Margin Level to determine the potential dangers of trading with high leverage.

EXAMPLE 13

Trader A deposited \$1,000 and started trading with a margin percentage of 100 percent. Trader B also deposited \$1,000 but started trading with a margin percentage of only 1 percent. Both traded a single standard lot of spot USDJPY and both lost 100 pips on the trade. This means that both traders lost their entire trading capital of \$1,000, in spite of the fact that Trader A employed no leverage whatsoever, unlike Trader B, who was exposed to an extremely high leverage of 100. High leverage had nothing to do with the losses. As you will see, the real deciding factor is money management. Regardless of leverage, both traders should have traded with an appropriate level of risk.

If Trader A maintains his \$1,000 deposit and margin percentage of 100, that trader will need an initial margin of \$1,000 to trade a single standard lot. Trader B started with a margin of only 1 percent, meaning

Trader B only requires an initial margin of $\$1,000 \times 0.01 = \$10$ to trade a single standard lot.

Now, each trader decides that the maximum lost per trade is five percent of the initial capital. Both Trader A and B were stopped out with a loss of only $\$1,000 \times 0.05 = \$20$ each. Both traders risked and lost equal amounts even though they were trading with very different amounts of leverage.

The problem begins when traders base the size of their positions on the maximum amount of positions affordable on margin, rather than on the maximum amount of risk tolerable. Under maximum affordability, Trader B would have bought $\$1,000/\$10 = 100$ standard lots, with the risk of losing all trading capital in a single pip move. Now that is dangerous! As we have seen, Trader B has the choice to go for maximum positions affordable or maximum risk affordable. It has nothing to do with leverage.

Therefore, high leverage is not dangerous unless you base your exposure on maximum positions affordable. Leverage is only "dangerous" when traders are under-capitalized and fail to use any form of risk control. The effects of maximum positional affordability are more damaging as useable margin approaches or falls below the initial margin level. And that's the whole story!

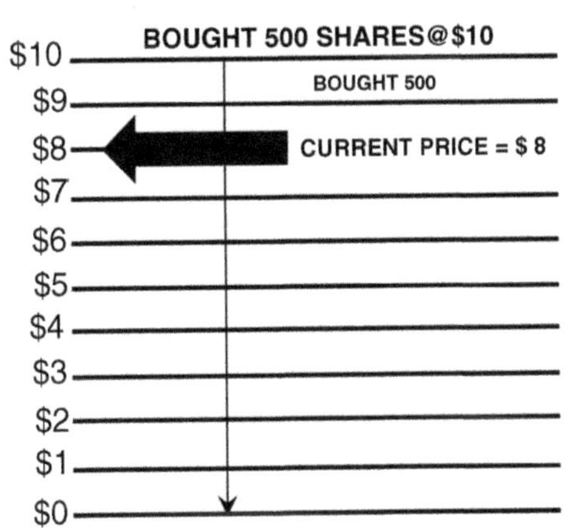

FIGURE 5.6 Finding the CMV

CALCULATING THE INITIAL AND CURRENT MARKET VALUE

Initial Market Value (IMV) = Total shares × Entry Price (EP)

Current Market Value (CMV) = Total shares × Current Price (CP)

Since (IMV/EP) = total shares, it follows that CMV = (IMV × CP/EP). Therefore, the current market value of all the shares invested is inversely proportional to BLR.

For multiple unhedged positions, CMV = Total Lots × Average Price × (CP/EP).

CMV = CP × Total Lots = \$8 × 500 = \$4,000 in Figure 5.6.

Similarly, CMV = IMV × (CP/EP) = (50 × \$10) × (\$8/\$10) = \$4,000.

CHAPTER 6

The Mechanics of Securitization and Monetization

In order to trade, you have to continually securitize money and monetize securities. In vibratrading, to monetize a security means to exit a position to lock in a profit, or in other words, to realize profits by converting them into cash. To securitize money, or equity, means to enter new long positions by converting cash into shares. In the process, we expose our account's equity to directional risk.

There are essentially three ways to monetize securities:

1. Monetize an arbitrary dollar amount.
2. Monetize IMV.
3. Monetize profit.

EXAMPLE 1

Monetizing an arbitrary dollar amount, as in Figure 6.1, reduces exposure in the markets. Assume we bought 300 shares at \$6, and that IMV is \$1,800 and CP is \$8.

$$\text{Therefore, CMV} = \$1{,}800 \times (8/6) = \$2{,}400.$$

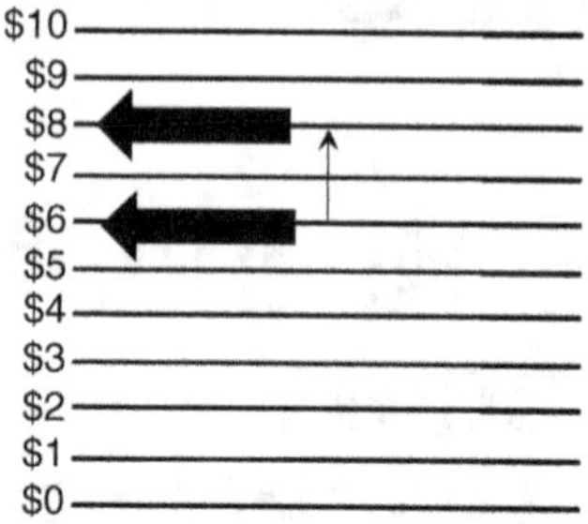

FIGURE 6.1 Increase in CMV

If we needed to monetize $1,200, that is, to remove half, we would use the formula below.

$$\text{Shares to Sell} = \frac{\text{ARB. \$Amount}}{\text{Current Price}}$$

By dividing amount by current price, $1,200/$8 = 150 shares to sell.

Example 2

We monetize IMV for a break-even exit. As in Figure 6.1, assume that we bought 300 lots at $6. Again, IMV is $1,800 and CP= $8.

$$\text{CMV} = \$1{,}800 \times (8/6) = \$2{,}400.$$

To figure out how many shares to sell in this situation to remove all IMV, we use the equation shown below.

$$\$1{,}800/\$8 = 225 \text{ shares.}$$

$$\text{Shares to Sell} = \frac{\text{IMV}}{\text{Current Price}}$$

Hence, we sell off 225 shares to remove all invested risk, while leaving the balance of 75 shares for a potentially risk free profit. We have in effect removed all initial capital investment and achieved a break-even exit.

EXAMPLE 3

To monetize, or lock in, profit in that same scenario, we would use our previous calculations to find the CMV and IMV and then use the equation given here.

IMV = \$1,800
CP = \$8
CMV = \$1,800 × (8/6) = \$2,400
\$Profit = CMV − IMV = \$600
Shares to Sell = \$600/\$8 = 75 shares.

$$\text{Shares to Sell} = \frac{\$\text{Profit}}{\text{Current Price}}$$

MONETIZING IN MARGIN AND NON-MARGIN ACCOUNTS

There are significant differences between margin and non-margin trading in terms of risk and extracting profits. Later chapters cover the distinction in more detail. However, you can immediately see it is possible to extract the full profit from positions in a non-margin account without exiting all the positions, whereas in margin accounts we are required to exit all positions to extract the full profit. Figures 6.2 and 6.3 illustrate this concept.

In the case of realizing, or monetizing, the original risk, both margin and non-margin traded accounts allow the removal of the initial investment without closing all positions, thereby rendering the trade risk free (see Figures 6.4 and 6.5.)

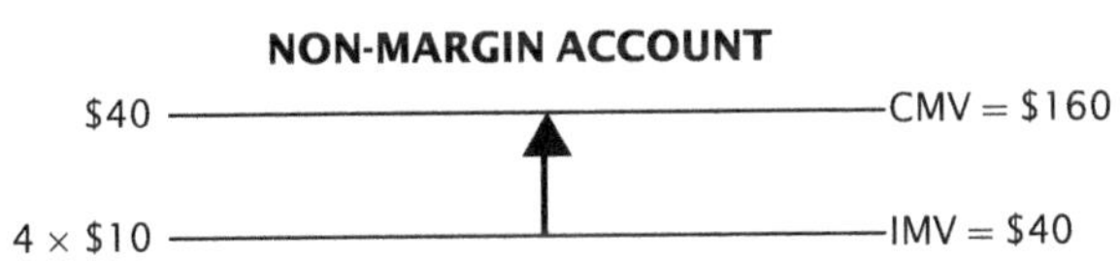

FIGURE 6.2 Non-margin account

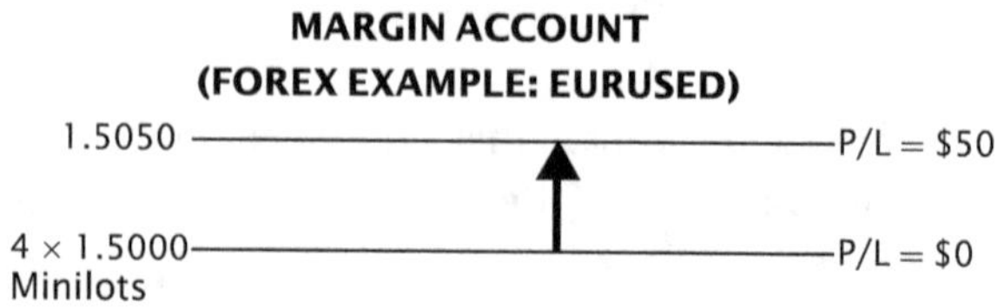

FIGURE 6.3 Margin Account

To extract (monetize) the full profit, we need to liquidate:

$$\begin{aligned}\text{Sell} &= \text{Profit/CP} = (\text{CMV} - \text{IMV})/\$40 \\ &= (\$160 - \$40)/40 = \$120/\$40 \\ &= 3\ \text{Shares}\end{aligned}$$

In non-margin trading, we can extract full profit without liquidating the entire position size. This is a significant advantage.

To extract (monetize) the full profit, we need to liquidate:

$$\begin{aligned}\text{Sell} &= \frac{\text{Profit Required}}{\text{Actual Profit}} \times \text{Lotsize} \\ &= (\$200/\$200) \times 4 \\ &= 4\ \text{Minilots (i.e., must liquidate all)}\end{aligned}$$

In margin trading, we cannot extract full profit without liquidating the entire position size. This is a limitation.

To summarize, the only difference between margin and non-margin trading, as far as monetizing is concerned, is the ability to extract full profit and risk in non-margin trading without exiting all positions. The extraction of full initial risk, which essentially creates a risk-free or break-even trade, is identical in margin and non-margin trading.

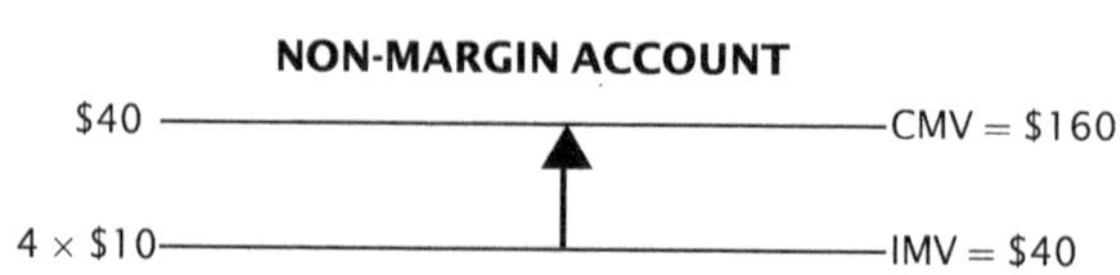

FIGURE 6.4 Monetizing Full IMV

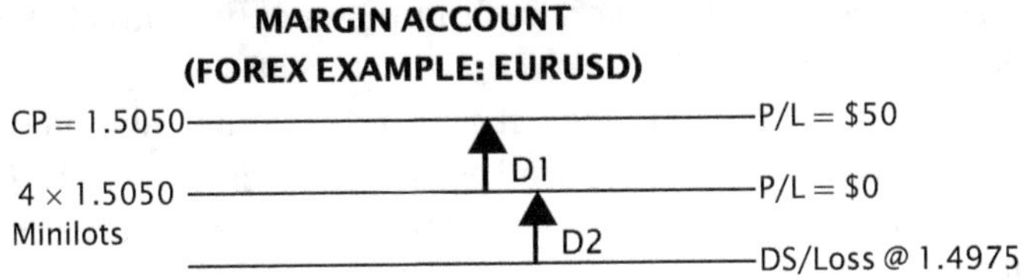

FIGURE 6.5 Monetizing Full Risk

To extract (monetize) the full IMV, we need to liquidate:

$$\begin{aligned}\text{Sell} = \$\text{IMV}/\text{CP} &= \$40/\$40\\ &= 1 \text{ Share}\end{aligned}$$

or:

(IMV/CMV) × Shares = (40/160) × 4 = 1 Share.

In non-margin trading, we can extract full initial risk.

To extract (monetize) the full risk, we need to liquidate:

$$\begin{aligned}\text{B/Even Exit} &= \frac{\text{D2}}{\text{D1} + \text{D2}} \times \text{Lotsize}\\ &= (\$25/\$75) \times 4 \text{ Minilots}\\ &= 1.33 \text{ Minilots (i.e., A B/Even Exit)}.\end{aligned}$$

Similarly, in margin trading, we also can extract full initial risk.

SECURITIZING PROFITS AND RISK CAPITAL

There are essentially two ways of securitizing capital: securitizing profit and securitizing risk capital.

Securitizing profit entails putting your profit to work. For example, if you made an earlier profit of $2,400 with a CP of $8, up from an initial $6, you could use the formula below to determine the number of shares to be sold to risk that profit.

$$\text{Shares to Buy} = \frac{\$\text{Profit}}{\text{Current Price}}$$

$$\$2{,}400/\$8 = 300 \text{ shares.}$$

To securitize, or convert cash in the account into shares, we simply divide the amount to be risked by the current share price.

On the other hand, securitizing risk capital involves putting risk equity to work. Say the required equity to be risked was \$5,000. If the CP is \$8, you can securitize equity by selling 625 shares, as illustrated below.

$$\text{Shares to Buy} = \frac{\$\text{Risk Equity}}{\text{Current Price}}$$

$$\$5{,}000/\$8 = 625 \text{ shares.}$$

Now that we are familiar with the process of monetizing and securitizing, we can analyze some simple scaling constructs.

THE BASIC PRINCIPLES OF PRICE ACTION

In order to grasp the concept of vibratrading, you first need a solid understanding of basic price action principles. This chapter presents a review of the main principles that govern the interaction between single, multiple, hedged, and unhedged positions. It also covers many of the preliminaries required in later chapters.

Unhedged Price Action Principles

Unhedged positions are either all longs or all shorts. Multiple mixed positions of longs and shorts will be discussed under Hedged Price Action Principles.

> *Two equal and opposite orders at the same price level will be equivalent to a null position, with no directional risk or overall net long or short exposure.*
>
> *Principle of the Null Positional*

In other words, holding a long and short open position at the same price level is similar to having no position held. Basically the longs and shorts will square each other off, effectively resulting in a zero net position. This is

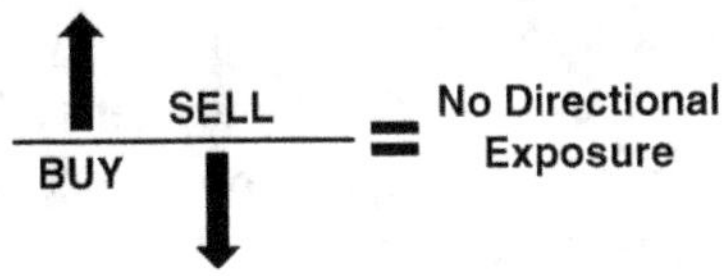

FIGURE 6.6 A Null Positional

true if, and only if, they are of equal and opposite positions. Otherwise, there will be a net long or short directional exposure. The principle of the null positional is illustrated in Figure 6.6.

Another implication of this principle is that executing a sell to close followed by a buy to open at the same level is redundant and essentially resolves into a sell and buy, with no effective change in net directional exposure. (This is why there is no Unbounded Upside Microsiso mechanism.) Such action serves no advantage whatsoever and will incur additional and unnecessary trading costs (see Figure 6.7).

Effects of Negative Spread Bias on Short and Long Entries

It is vital that the trader understands the negative effect of spread bias as this will affect the construction of more "sensitive" scaling mechanisms in less liquid markets. Many novice traders trade at prices that they believe represent what they see on the charts, failing to take into account spread bias.

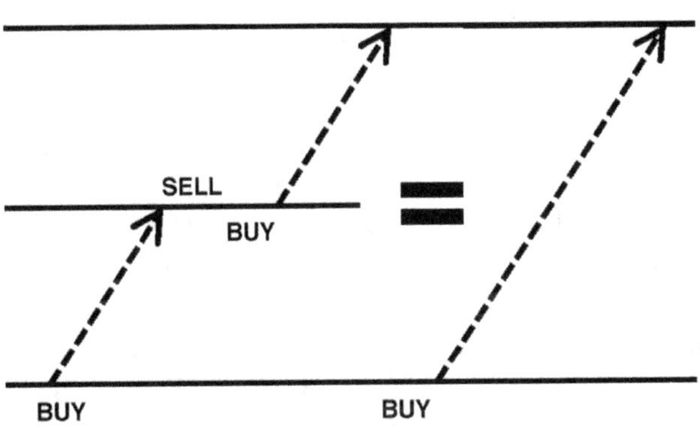

FIGURE 6.7 Initiating Null Positionals in a Trend

The issue of negative spread bias arises in only one situation: when a trader enters long orders based on visual cues derived from chart analysis. Basically, the spread represents the difference between the bid and ask prices (the buy and sell price quotes).

Spread bias has a negative effect on:

1. Market orders
2. Working or pending orders
3. The Reward to Risk Ratio.

Before proceeding, let me remind the trader that most price charts are bid-based, as opposed to being based on the ask price. The actual price pattern as seen on candlestick, line, and other charts has to be translated vertically upwards by the exact amount of spread in order to reflect an equivalent ask-based price chart. Therein lays the discrepancy.

A few points regarding market orders and pending orders:

1. All sell/short orders, both market and pending, will be filled at the trader's desired price level per the charts.
2. All long market orders may be filled at a price not anticipated by the novice trader, as the price level at which they are filled is not reflected directly in bid-based charting. This gives rise to price disadvantages.
3. All long pending orders are filled at the exact price anticipated, but it still presents the opportunity for timing disadvantages.

Let us refer to a few examples to illustrate these very important points.

Example 1

Negative spread bias does not affect shorts getting filled at desired prices, since the charts are bid-based. So, assume you want to short a price pattern called the double top, as seen in Figure 6.8. As you can see, the spread is \$1.20 – \$1.00, or \$0.20. You enter a market order wishing to short the market at the very moment price hits the \$1 level, thus profiting from a potential double top price reversal to the downside, as seen from the left side of the illustration below. You hit the bid and entered a short at \$1 without any problems.

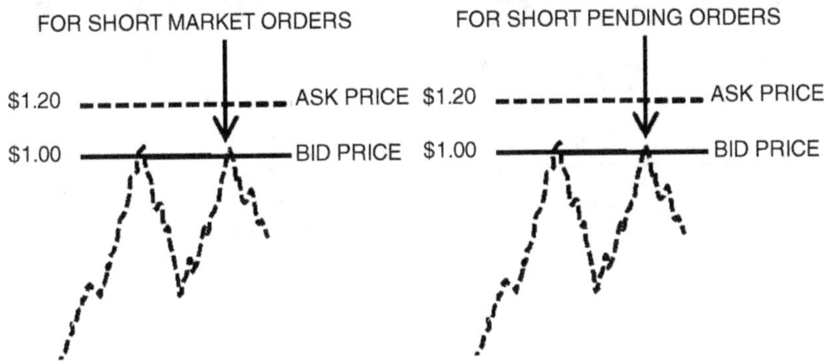

FIGURE 6.8 Shorting a Potential Double Top

Example 2

Let's now assume that you placed a pending order. Again, you wish to short the double top. You enter a pending order to short once price hits the $1 level, hoping to profit from the same double top price reversal, as seen on the right hand side of the illustration. Your bid order was filled at $1 later that day. Again, there are no problems encountered.

As you can see, there is no discrepancy in the fills, since the trader selected a price based on the bid prices, which is exactly what is reflected on the charts and the quote screen. In a nutshell, what you see (or think you see), is what you get. However, spread bias does have a negative effect on long orders, as illustrated in the next example.

Example 3

Let us assume that you placed a market order to long, as well as a buy stop entry order (pending order). The spread is still $0.20.

Figure 6.9 depicts a long market order (on the left) and a long pending order (on the right). We see that if we enter a market order to long a breakout at five cents above the $1 price level, which was the previous swing high point indicated in the chart, we will in fact be long at $1.05 + $.20 or $1.25. Even though we based our breakout on the charts, we will never actually long at the indicated price level of $1.05 due to the spreads. This price disadvantage is known as expensive longs, and no trader can overcome this built-in cost. Even if the trader accounts for spread bias and adjusts to enter at

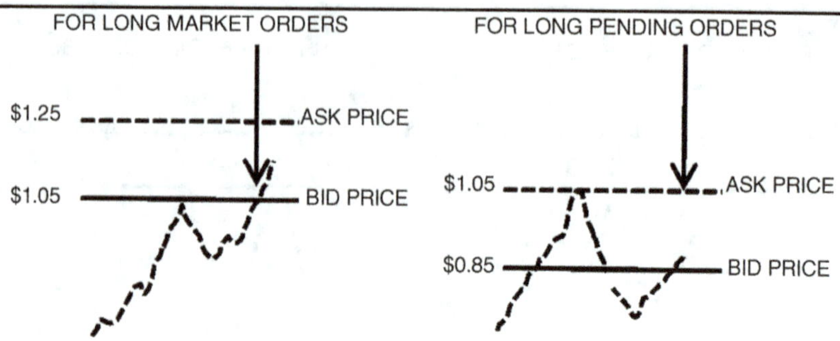

FIGURE 6.9 To Long an Upside Breakout

exactly $1.05, this gives rise to another problem: early longs. We see this problem on the right hand side of Figure 6.9. Here, the trader is looking to initiate a pending order to long once the price hits five cents above the $1 level, which was the chart-based level of the previous swing high point. As the price begins to rise, the buy stop entry is filled at $1.05, exactly as intended. Unfortunately, upon consulting the charts at this point, the trader will see that there is no visible sign of any breakout! The trader has inadvertently entered the long position too early. This is a timing disadvantage, since entering before a breakout is confirmed is a higher risk, aggressive entry. For all practical purposes, price could easily reverse if it fails to penetrate the prior swing high.

Example 4

What if you placed a pending order to long? In this example we will study the negative effects that spread bias has on buy limit pending orders. Refer to Figure 6.10. (Spread is $0.20.)

As seen from the illustration, the trader intends to enter a long position once price tests the prior swing low point, hoping for a reverse to the upside. The trader therefore places a buy limit order at the prior swing low point of $1. The limit order is eventually filled but, to the novice trader's utter surprise, the chart is already indicating a breakout to the downside! This, again, is caused by the fact that the bid price, which "draws" the charts, is situated below the ask price. Hence, once the ask price has tested the $1 level, the actual chart will show price at the bid price of $0.80 instead.

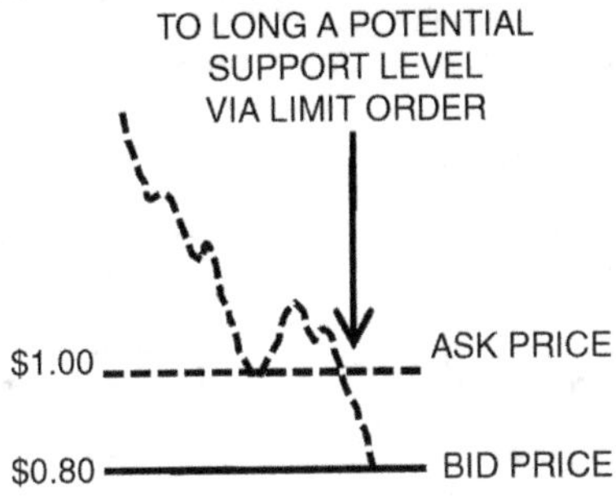

FIGURE 6.10 To Long a Potential Support Level

As a result, the trader gets into a reversal trade after a breakout has been confirmed on the charts. Of course, the advanced trader may account for this by re-adjusting entry to $1.20, which accommodates the spread bias, and in the process get in at the intended, but more expensive, price.

This is called the problem of "late longs."

THE EFFECTS OF NEGATIVE SPREAD BIAS ON REWARD TO RISK RATIO

To understand the Reward to Risk Ratio (RRR) effects of negative spread bias on long entry, let's focus on the left side of Figure 6.11. Imagine you enter a long position at a price level of $1. The spread is $0.20. You then enter a long position the very moment the chart indicates a breakout at $1. In fact, you have entered this long position at $1.20 without realizing it, by failing to take into account that the ask price, the buying price, is $0.20 above the breakout price indicated in the charts. You now punch in the stop loss at $0.50 and a profit target at $1.50, assuming the RRR is $0.50/$0.50 or 1:1.

Unfortunately, since the actual long entry was at $1.20, the real potential profit is only $0.30 ($1.50 – $1.20), and not $0.50 as assumed. Furthermore, the actual risk involved in this trade, should price trigger the stop loss at $0.50, will be $1.20 – $0.50, or $0.70. The RRR is actually $0.30/$0.70 (0.43:1) and not 1:1 as initially assumed, a clear example of negative spread bias affecting the RRR of long entries based on chart prices.

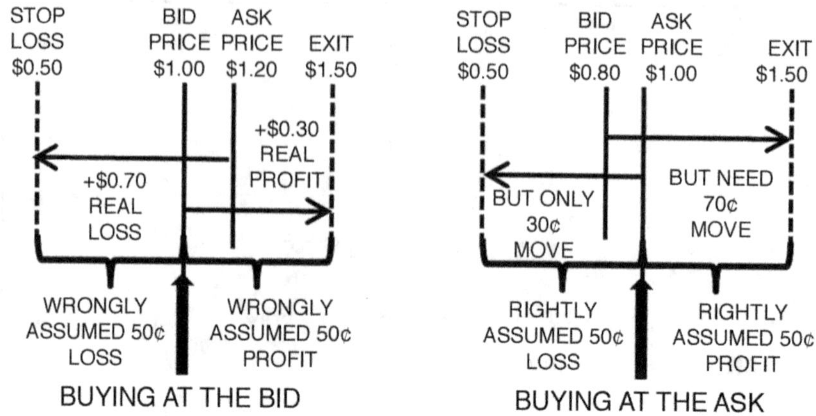

FIGURE 6.11 Negative Spread Bias Affecting the Reward to Risk Ratio

We now refer to the right hand side of the illustration above. Here, the trader has entered a long position by virtue of a pending order at $1, and has also punched in the stop loss at $0.50 and the profit exit at $1.50. (Note that it does not matter whether the entry was made via a buy stop or buy limit order as the RRR will be affected in the same way.) The novice trader assumes that the RRR is 1:1. But is this really the case?

The potential profit will be the sell price less the buy price.

$$\$1.50 - \$1.00 = \$0.50.$$

The risk involved is just the stop price less the buy price.

$$\$0.50 - \$1.00 = - \$0.50.$$

So, the final RRR is $0.50/$0.50 or 1:1.

This does not mean that the spread has no negative effects, but rather that they are somewhat less obvious to the novice trader. The negative effects are manifested in a different manner.

Upon further examination, we see that in order to make a reward of $0.50, price has to move by an amount of (Sell Price − Buy Price + Spread) $0.70. Conversely, to get stopped out, price only needs to move 30 cents instead of the 50 cents originally assumed. Note that spread bias also affects trades that short at the bid or ask.

> *There is an inherent negative bias built into every trade in the form of the spread, which creates a resistance toward profitability and relative ease toward loss.*

This, in essence, is what the term "negative spread bias" means. The existence of the spread causes either diminished RRR or increased resistance to profit accompanied by a bias toward losses. It does not matter if the trades were entered long or short, as all trades will suffer the effects of negative spread bias.

In summary:

1. Long market orders result in a price disadvantage due to negative spread bias if upside breakouts are based on bid prices, and may result in a timing disadvantage if upside breakouts are based on ask prices.
2. All long pending orders result in a timing disadvantage.
3. Negative spread bias affects the RRR for both long and short trades with accompanying stops and profit targets.

Average Pricing on Multiple Open Unhedged Positions

To get started as a vibration trader, you must fully comprehend price averaging of multiple open positions.

In the following illustration, we initiate a series of long positions. We bought fifteen shares at $10, ten at $9, eight at $7, and seven shares at $2. We can represent this as an aggregate share size at the breakeven or average price. To obtain the average price, we add up the market value of all the shares and divide it by the total number of shares, as shown in Figure 6.12.

In the event of having multiple open long or short positions, there will exist an average price representing all positions that can be effectively collapsed into one aggregated share size at that price.

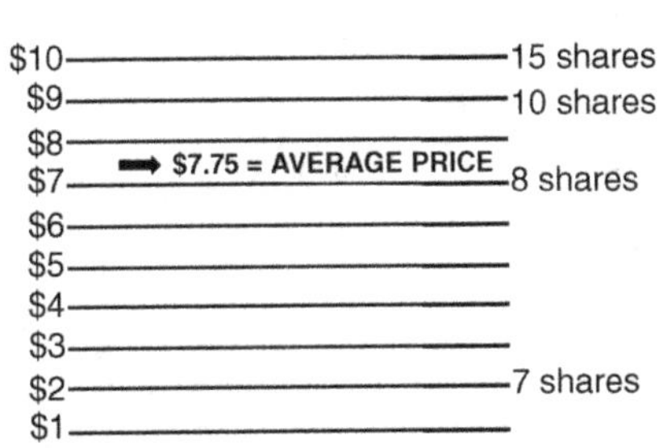

FIGURE 6.12 Finding the Average Price
Note: Average Price is also called Break-Even Price (B/E).

$$\begin{aligned}\text{Av. Price} &= \frac{\text{Total Market Value}}{\text{Total Shares}}\\ &= \frac{\$150 + \$90 + \$56 + \$14}{\text{40 Shares}}\\ &= \$7.75\end{aligned}$$

Therefore, instead of treating each position separately, we can reduce multiple positions into a single representative and equivalent position size at the average price, simplifying the entire position. In this example, we may view our entire position as being long 40 shares at \$7.75. One advantage of price averaging is that it reduces the complexities involved when calculating the profit and loss for multiple positions, as evidenced by the next example on profit and loss calculation.

Let us assume that price now rallies to \$10. Since we know our average price is \$7.75, it is remarkably simple to calculate the profit made by the move in price. From the illustration in Figure 6.13, we see that all we need to do is find the difference between the current price (CP) and average price (AP), and multiply that by the total position size of all the shares owned.

We represent the multiple positions as 40 shares at \$7.5.

Therefore, calculating P/L is now a simple matter:

If price is now at \$10, what is the P/L?

$$\begin{aligned}\text{P/L} &= \text{Total Sharesize} \times (\text{Current} - \text{Av. Price})\\ &= 40 \times \$2.25\\ &= \$\,90.\end{aligned}$$

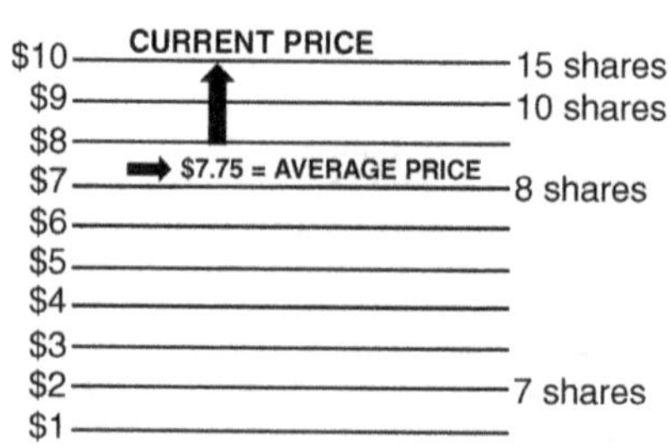

FIGURE 6.13 Using Average Price to Find P/L
Note: Same Principle applies to multiple short positions.

Calculating each position separately also yields:

$$\$\,10 + \$\,(8 \times 3) + \$\,(7 \times 8) = \$\,90$$

If we calculated the profit by adding up the individual profits made by the various positions, we still get $90 as the final answer. Therefore, average price is extremely useful in simplifying calculations in complex, multiple position portfolios.

The vibration trader must also know how to lock in profits, especially in multiple, open unhedged positions.

Let us assume that our average price is $7.75, as in earlier examples, and that the current price is $10. We know from the previous example that our profit is $90. To lock in this profit, i.e., remove it from any further directional risk, we monetize it and keep it in the form of cash in the account.

Divide the profit by the current price to find the number of shares to liquidate, or monetize, in order to lock in the profit. In this example, the vibration trader must liquidate nine shares to lock in the $90 profit (see Figure 6.14.)

If current price is $10, and profit = $90, to monetize profit we need to liquidate:

$$\begin{aligned} &= \$90/\text{Current Price} \\ &= \$90/\$10 \\ &= 9\ \text{Shares} \end{aligned}$$

Shares have been used throughout the examples but strictly speaking, it should be shares for equity traders and contracts for futures traders.

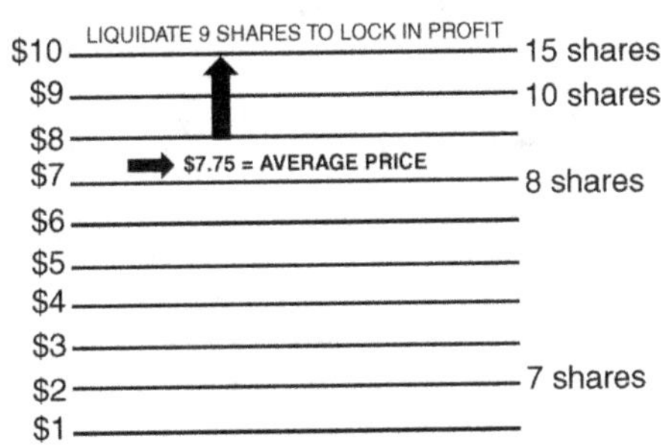

FIGURE 6.14 Monetizing Profit
Note: Monetizing Profit means locking-in profit.

Break-Even Exits on Multiple Open Unhedged Positions

There may be instances in which the vibration trader needs to exit all except riskless positions, that is, shares bought on profit. In essence, the trader is removing all positions from any further market exposure, while leaving the risk free shares to earn additional profit. In order to break even, we must monetize the value of the IMV.

We need to divide the IMV by the current price in order to find out how many shares to liquidate in order to execute a break-even exit. From Figure 6.15, we see that we needed to liquidate 31 shares. This leaves nine risk free shares that may be left in the market to potentially generate additional profits.

Calculating the CMV using Average Price:

$$\begin{aligned} \text{CMV} &= \text{IMV (In multiple form)} \\ &= \text{Average Price} \times \text{Total Shares} \\ &= \$7.75 \times 40 = \$310 \end{aligned}$$

Monetizing IMV (or CMV):
If current price at the rate \$10, To monetize IMV We sell

$$\begin{aligned} &= \text{IMV/Current Price} \\ &= \$310/\$10 \\ &= 31 \text{ Lots/Shares.} \end{aligned}$$

By doing this we have effectively executed a break-even exit, leaving only securitized profits in the market for riskless trading.

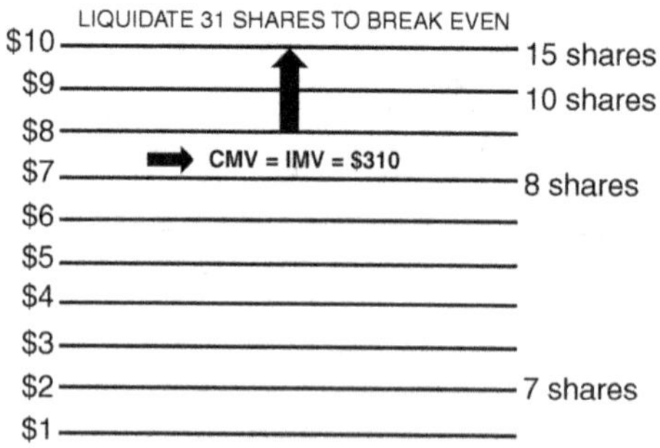

FIGURE 6.15 Liquidating at Break-Even
Note: Total Monetized IMV and profit must add up to (9 + 31) 40 shares.

HEDGED PRICE ACTION PRINCIPLES

Multiple hedged positions consist of both long and short positions. Learning these hedged price action principles will prepare the vibration trader for a better appreciation of vibrational strategies like Macrosisosi Vibrahedging, Short Rider, and Short Scaler.

Hedge Creation Costs

From the Principle of the Null Positional, we saw that we could simulate a null position by holding an equal number of open buys and sells at the same price. The price to create a hedge is called the spread cost, and does not include commissions, interest charges, or other fees. Note that the equal and opposite buy/sells are hedged directionally, but trading costs are not hedged.

In Figure 6.16, the number of both long and short shares (which is always equal) is denoted by N. Let N equals to 2,000 shares and the spread be $0.10.

$$\begin{aligned}\text{Therefore, the spread cost} &= (\text{long} + \text{short shares}) \times \text{spread}\\ &= (\text{N} + \text{N}) \times \text{spread}\\ &= 4,000 \times \$0.10\\ &= \$400\end{aligned}$$

Net Position Bias and Profit/Loss Path Independence

A set of multiple hedged open orders is biased in the direction of the larger position size. So, if we have ten open longs and nine open shorts, we are

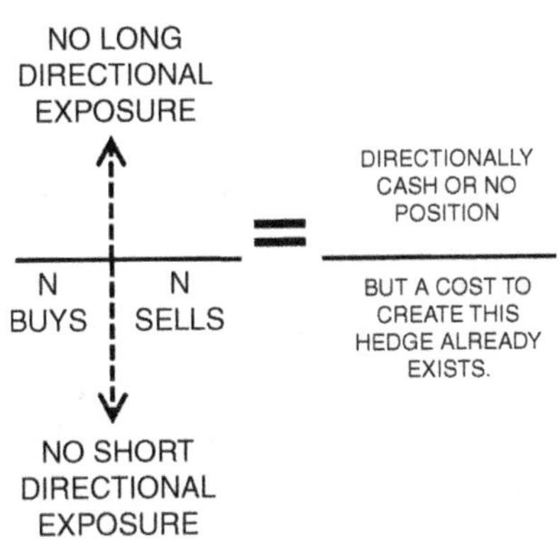

FIGURE 6.16 The Perfect Hedge

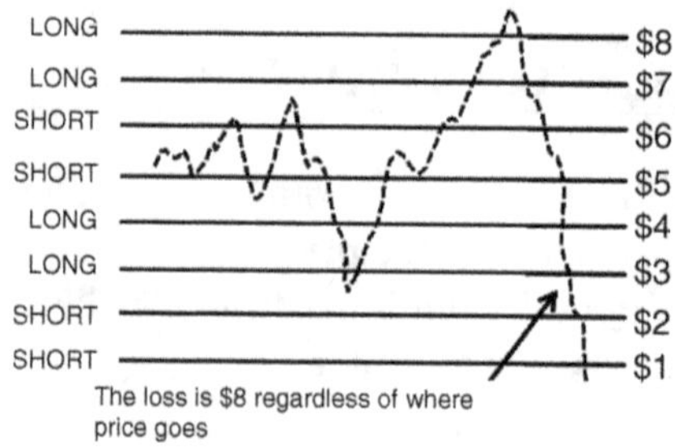

FIGURE 6.17 Path Independence
Note: The Fixed P/L is −8 irrespective of where price goes.

long biased. We will profit as price rises and lose as price falls since we have a net long exposure.

If the number of open long positions equals the number of open shorts, then the profit and loss is said to be price independent. The overall profit or loss is fixed no matter which direction or path the price takes, or how far price extends.

In Figure 6.17, let us assume that we go long and short, one share each, at the various indicated price levels. We see that, once all the shares are in open positions, the path that price takes is totally irrelevant as the P/L is fixed. In this example, we have an overall loss of $8. It does not matter if price rallies or falls, the loss will remain as $8 until the positions are squared.

The manner and order in which long and short pairs are squared off is also irrelevant. It does not matter if the long at $8 is squared off with the short at $6, $2, or $1—the resulting overall P/L is still the same. Since this squaring off essentially locks in the profit or loss, and is price independent, the trader has actually secured profit, or on the flip side, locked in all the risks involved.

However, this locking in of the P/L still represents an unrealized P/L, as the long and short positions are still open in the market. We therefore refer to this type of locking in as soft locking.

Each pair of equal and opposite size and position will soft lock in either profit or loss, and will be independent of all future price action. In addition, each pair that squares off contributes to the overall P/L.

As you can see from Figure 6.18, a position may be squared off with the adjacent opposite order or with one that is at another price level. The overall P/L is soft-locked, regardless.

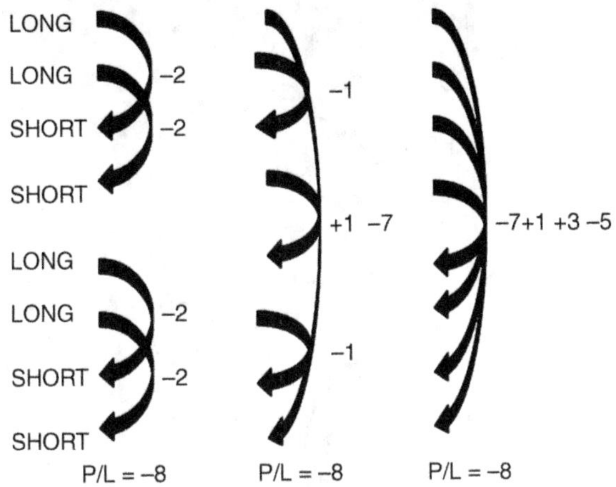

FIGURE 6.18 Soft-Locking P/L
Note: The P/L is fixed in the figure above irrespective of where price goes.

Hedged P/L Calculation

Any position left unhedged will give rise to some directional exposure, and as a result a running P/L. The running P/L is found by multiplying the number of unhedged shares left by the difference between the current price and the entry price.

The total running P/L will be the sum of the unhedged running P/L and the total soft-locked P/L, as seen in Figure 6.19. In that example, we have one long position that is unhedged, that is, not squared off with an equal and opposite position.

The remaining unpaired orders which will either be all longs or all shorts will determine the running P/L.

Running P/L is price dependant.

$$\begin{aligned} \text{Running P/L} &= \text{Balance Unpaired Share Size} \times (\text{CP} - \text{EP}) \\ \text{Total P/L} &= \text{Softlocked P/L} + \text{Running P/L}. \end{aligned}$$

Having established the basic principles of price action and positional analysis, we shall now turn our attention to the concept of boundedness and its role in vibratrading.

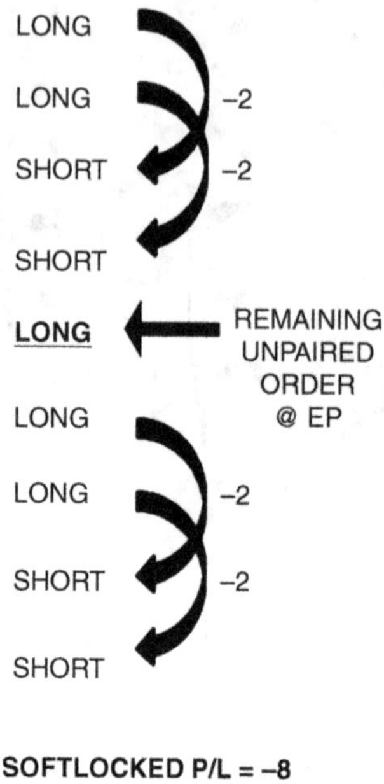

FIGURE 6.19 Directional Exposure of Unpaired Orders

CHAPTER 7

The Principles of Boundedness

It is universally acknowledged that the majority of directional traders lose sooner or later. In the short term, the lucky may experience some level of profitability. As stated in the introduction, loss in the long term is mostly attributed to the single most damaging mechanism ever introduced to directional trading: the stop loss.

A depletion of equity or capital bleed is experienced every time a stop loss is hit. The constant application of stop losses for short term protection reverts, functionally-speaking, into a "stop profit" mechanism instead. If the rate of income is less than the rate of loss, there will come a time when the account will inevitably be obliterated. Normally, the profit preclusion effects of a stop loss overwhelm the shorter term profit protection. Traders have absolutely no control over the win-loss percentage and distribution, which are determined by market action alone. Expectancy itself is, irrefutably, a non-controllable or non-actionable performance parameter in light of those effects. Hence, in the majority of cases, expectancy is "expecting" the win-loss percentage to remain higher than the expected value in order to remain positive. This is highly unlikely as the market, being mostly unpredictable, will express itself in all manners imaginable and in the process result in the unavoidable mismatch between methodology and market action. When such mismatch occurs, the account will experience periods of drawdown.

If the drawdown is severe, protracted, and regular as compared to minor or infrequent run-ups in equity, then it will just be a matter of time before the equity is totally depleted.

Let me illustrate this with a simple example. Say we have a consistent winning percentage of 40 percent, and that we have set our risk to reward ratio at 1:2. We "expect" the winning percentage to remain at 40 percent. This means that for each dollar we risk, we make two dollars. Hence, the expectancy of the trading system for each window of 10 trades is:

$$\begin{aligned}\text{Average Profit/Loss per 10 Trades} &= (0.4 \times \$2) - (0.6 \times \$1)\\ &= +\$0.20 \text{ per Trade.}\end{aligned}$$

Since there is an average profit, we say that expectancy is positive. Now, consider if there were some market irregularity or mismatch during the first 10 trades and we did not manage to achieve the 40 percent winning percentage. What if we lost all 10 trades and the market then decides to resume its "expected" behavior for the subsequent trades?

Assume each trade window is 10 trades.

Total loss in previous window is −$10.

Total Expected Profit per window is $2.

Therefore, the number of windows required to overcome the loss is 5 windows ($10 / $2) or 50 trades.

To overcome our inability to achieve the four required winning trades during the first window, we will now be required to make a total of 50 trades just to break even. And that is not even the worst case scenario. If the trader is required to make 50 more trades just to break even, then, by virtue of the sheer number of trades required, the probability of such an incident recurring is substantially higher. Once the trader starts down this slippery slope, there is almost no chance of recovery. The losses literally start to spiral out of control. The only way out is to stop trading altogether, since topping up the trading account is pointless and does not solve the problem.

Maintaining Potential Earning Power: Profit Drivers

So, we need to implement a trading methodology that avoids all forms of capital depletion. This means that the trader must not employ any trade mechanism that depletes capital permanently. By permanent depletion

I mean the potential inability of an account to maintain its original open positions during drawdown. If you do not exit any trades in loss, the drawdown does not affect the future earning potential of the trading account. The number of open positions, or profit drivers, remains intact.

Let's say Trader A allocates five percent of his initial trading capital, or equity, per trade. Each time his stop loss is hit, he loses five percent of his capital. If he continues to experience loss, there will come a time when he will be unable to open any more positions due to a lack of capital. This loss is permanent. There is no repair strategy, except to fund the account again.

Now let us consider another scenario. Trader B allocates five percent of capital to buy some shares at $50. She has now some open positions, and is directionally exposed. As the price of the shares fall, she decides to continue scaling in more shares, allocating five percent for every $10 decline in price. Not only has she added some shares to the initial amount, but she has bought increasingly more shares at each successively lower price. Hence, as the price of the shares decline, and as she continues scaling in, her earning potential will increase dramatically. She has, effectively, more profit drivers than when she started, even though she has also exhausted all her useable trading capital.

The significant difference is that Trader A has no potential future earning power, whereas Trader B is open to potential profit as she is still holding a large number of open positions. She has the maximum amount of profit drivers possible, whereas Trader A has none whatsoever. Therefore, should price now move in Trader B's favor, she has the opportunity to break even, if not profit. In fact, Trader B has the opportunity to continue extracting vibrational returns at that current price.

Trader A lost his entire account from capital bleed. Trader B has no extra cash in her account, and the total market value of her shares may be lower than the initial account equity, but she is nevertheless still holding every share she bought. Trader B is still exposed to future profits even if she does nothing further, whereas Trader A needs to fund the account in order to continue trading. From this perspective, Trader B is comparatively better off.

The idea of avoiding capital bleed is the very foundation of vibratrading. The whole idea is to be able to maintain future earning power by making sure that these profit drivers are maintained as long as possible.

CAPITAL BOUNDEDNESS

In order to keep profit drivers intact, we need to address the issue of capital. If Trader B were to instead average up, that is, scale in short positions as price rises, then the entire exercise of maintaining the profit drivers may be compromised. This is because there is no upside boundary in price, but there is a lower boundary, namely $0, or what I call Zero Test Level, or ZTL. This lower boundary is absolute and there is absolute agreement to its price. Therefore the vibrational trader is able to calculate the entire capital required to average down to ZTL. This maximum amount is called the total allocated capital. Of course, this capital is only allocated when price declines to the lowest buy level, as stipulated by the vibrational trader. The amount of capital required to open new positions or buy more shares is called the working capital. Hence the maximum working capital is the total allocated capital.

The trader must not run out of working capital. This could happen in two ways: via capital bleed from the use of protective stop losses, or from insufficient capital to sustain the drawdown, resulting in a margin call. Equity traders will not experience margin calls if the shares have already been paid in full, and not purchased on margin. Therefore the vibratrader must not average down unless there is adequate working capital to hold shares right down to ZTL, either on margin or otherwise.

But, if the vibratrader decides to average up by scaling in short positions, there is essentially no upper limit as to the amount of working capital required, since there is no upper price limit. Neither can the trader use a stop loss as this would deplete the working capital permanently. This means that vibrational traders must never enter short positions. Since there is no preservation of capital, we say that there is no capital boundedness.

Capital boundedness requires that the vibratrader refrain from using mechanisms that:

1. Deplete capital permanently
2. Endanger open long positions
3. Expose the trader to positions with no upper limit in working capital.

Therefore the bounded vibrational trader must not use stop losses, short positions, or long options. Long options require premiums to be paid upon initiating a position in either a call or a put, and if the trade fails to break even, let alone produce a profit, that premium is permanently removed from the account's equity. Therefore capital boundedness is not preserved. We will see later that short options used within the vibrational construct are, for all practical purposes, riskless.

The Differences between Capital Boundedness and Unboundedness

Capital bleed, a symptom of capital unboundedness, is most pronounced when using stop losses in:

- highly volatile markets,
- unpredictable and irregular price behavior, or
- under-optimized stop loss management systems.

See Figure 7.1. Even though prices finally come back to the starting price, you have lost all your capital by letting volatility bleed profits. This effect is part of time in market risk.

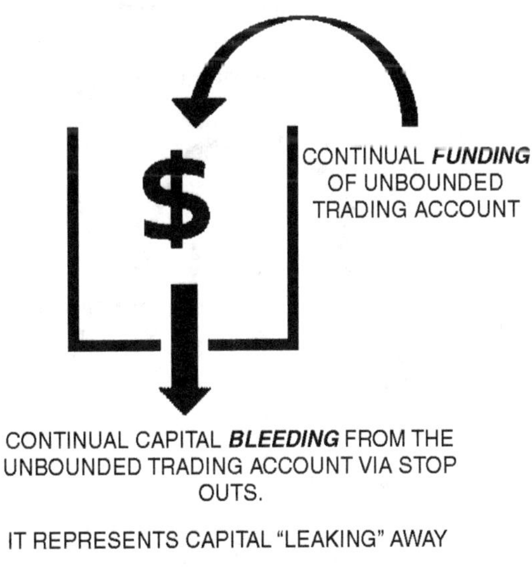

FIGURE 7.1 Unbounded Trading System

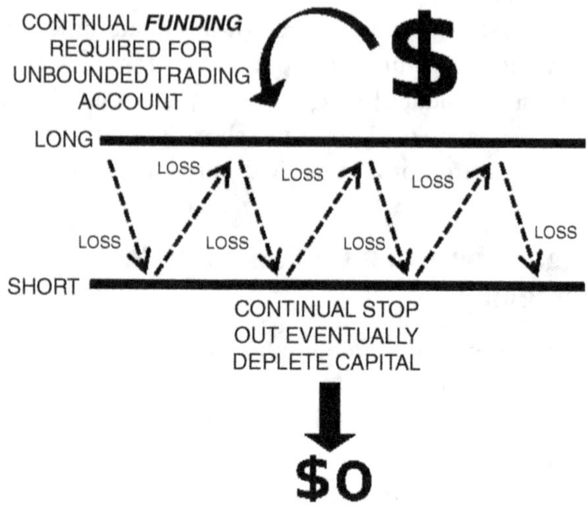

FIGURE 7.2 Capital Bleed from Stop Outs

With stop loss mechanisms, you must make sure that your rate of income is faster than your rate of loss, otherwise capital depletion is guaranteed (see Figure 7.2).

When trading with capital boundedness, we gain an edge. Profit is greatest when trading in highly volatile markets or unpredictable and changing price behavior. As time passes, the long-term edge of the vibrational trading methodology becomes evident, the exact opposite of directional trading. Figures 7.3 and 7.4 illustrate the essential differences.

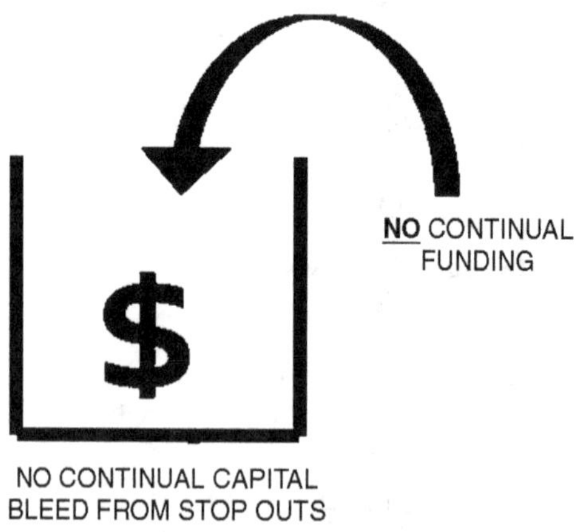

FIGURE 7.3 A Bounded Trading System

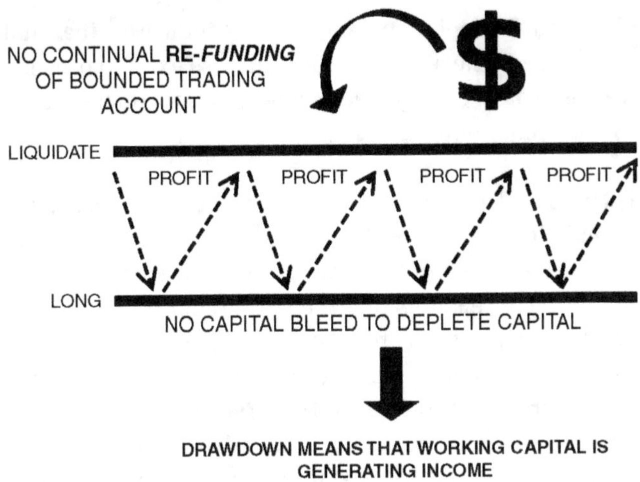

FIGURE 7.4 A Bounded Trading Environment

When price finally came back to the starting point, you had already profited on your capital just by letting volatility work. Time is working for you.

With these mechanisms, you do not need to ensure that your rate of income is greater than your rate of loss, since capital will not deplete. This is your never-ending income stream.

The Four Main Aspects of Capital Boundedness

The Preservation of Capital This preservation is intended to stem capital depletion or erosion over time that results from employing the use of stop losses. There must not be any positions that exit in loss. The balance curve must always point upward in vibratrading.

Capital boundedness, or preservation, is best described by the Initial Entry Rule:

> *If the final account's equity is less (or has the potential to be less) than the initial equity when price returns to the initial entry level, then that methodology is said to be unbounded.*

This means that if you want to find out whether a strategy is bounded, check to see if the final equity is less should price retest the original entry level. If it is, or has the potential to be, then that strategy is unbounded. Another way of determining capital boundedness is to check the balance curve. A downward sloping balance curve indicates that unbounded trading has occurred. The act of avoiding all loss exits to preserve capital directly gives rise to the other three facets of boundedness, namely order, directional, and range boundedness.

Type of Capital The only type of capital allowed under the principles of capital boundedness is disposable capital. We insist on capital that is not required elsewhere and therefore may be left to accumulate vibrational profit indefinitely. This ensures that the vibratrader is able to set aside the total allocated capital without redeeming the funds after the vibrational construct is implemented. Removing part of the allocated fund midway may damage the vibrational profits and give rise to unnecessary squaring of profit drivers. In directional trading, it is said that you must only trade with funds that you are prepared to lose. In vibratrading, you must trade with funds that you are prepared not to use.

Effective and Timely Deployment of Capital The disposable fund needs to be allocated effectively and with sufficient speed so as to avert any possibility of a margin call. Therefore, vibratraders are required to have the disposable capital ready for deployment at all times, if not necessarily in the account. Breaching this rule may introduce capital erosion, which is trading in an unbounded state.

Sufficiency of Bounded Capital Capital sufficiency means having enough disposable capital to tolerate a zero test level event, otherwise, this produces an intolerable lower trading range limit that will erode capital over time if breached. With the trading range bounded to actual price extremes, namely the ZTL on the lower extreme in price, the vibrational trader will be able to assign the maximum amount of allocated funds required by the type of vibrational construct chosen, as opposed to potentially unlimited funds in a "floating type" relative trading range where the price extremes are replaced by range extremes. Range extremes are only relative price extremes, and not absolute. Therefore stop losses get hit much more frequently than a security plummets to zero.

The Real Meaning of "Risk" in Vibratrading

In unbounded trading there is the possibility of capital depletion. This fact alone requires traders to balance the reward to risk equation to ensure longer term survivability. Reward to risk analysis does not work in the usual manner with bounded vibrational trading.

The idea is to win more than you lose on each trade. The loss comes from being stopped out. So to avoid being stopped out, you need to be able to time the market. Timing the market means to predict future price direction. Every time your prediction is incorrect, capital is depleted—a definite risk.

But in bounded trading, capital is never depleted. Even when the CMV of your investments is very low, your entire investment is working much more efficiently since PLR, BLR and MLR are higher.

> *Your entire position size is still functioning at full capacity generating returns.*

Boundedness and the Concept of Risk

This may be the most important paragraph in the entire book. It is critical to understand that in vibratrading, risk will only manifest itself if you allow yourself to fall under four conditions:

1. Not being able to fund long positions within a construct all the way down to the ZTL.
2. Not being able to extract vibrational returns between the entry price and the ZTL due to non-existent or incomplete downside constructs. We do not want any "dead zones" within the pyramidal structure. Without those returns, the vibratrader may never be able to reduce the cost basis of the initial investment. It is akin to a "buy and hold" scenario in which price never moves above the initial or average price.
3. Not being able to fund the account with capital that is disposable. We must allocate only capital that can be left to work indefinitely in return for vibrational income.
4. Entering naked or net short positions (unless it is part of an Upside Short Hedge approach).

If any of the four mistakes above are made, risk will manifest itself. It is imperative that you ensure that all four rules are adhered to in the strictest sense to avoid those situations. Otherwise, you will be exposed to "directional risk" and no more vibrational gains in your trading approach. Regardless of the vibradirectional or vibrabidirectional strategies used, they must be accompanied by constructs that abide to the four rules above.

If you abide strictly to the rules above, the only true risk in bounded trading is a ZTL event. As we know, for this to manifest equity ETFs must have all component stock prices go to zero, a near impossibility. We must remember that in bounded trading, "risk" is identified and expressed as running or working capital. There is no possibility of trading loss in bounded trading—only capital at work. This is a crucial difference.

DIRECTIONAL BOUNDEDNESS

We have seen that to protect capital boundedness, bounded vibratrading must only be long in a portfolio. Hence vibrational traders are bounded directionally. Shorts are not allowed within bounded vibrational constructs. We can still use various unbounded hedging and trend capture techniques that employ buy and sell stop orders to initiate and exit positions. However, there is no intentional use of these orders to establish a loss exit. They are unbounded as these orders are susceptible to negative slippage and gapping.

Directional boundedness is central to bounded vibrational trading. The basic mechanisms of the vibrational matrix itself must be oriented for oscillational profits, because vibrational traders never sell high and buy back lower. This means that vibrational traders buy to open a new position as price declines and only sell to exit as price rises. This set-up is required in order to extract vibrational profits. I covered in an earlier chapter that every time we buy as price falls it is called a downbuy, and every time we sell as price rises it is called an upsell. There are no upbuys or downsells in bounded vibrational trading as those break-out trades will suffer oscillational losses as price vibrates. Remember, all upsells are vibrational exits and all downbuys are vibrational entries.

In bounded vibratrading, the principle vibrational scaling mechanisms are:

1. **Macro/Microsiso:** The vibrational trader only executes downbuys and upsells, exiting either across a single interval (Microsiso) or across two or more intervals (Macrosiso).
2. **Upside Micro/Macrosiso**: The vibrational trader executes downbuys via buy limit orders as prices rise, and subsequently upsells higher across a single interval (Upside Microsiso) or across two or

more intervals (Upside Macrosiso). These mechanisms are bounded by the very fact that they are initiated below the apex, and hence have taken into account the total allocated capital required to maintain a vibrational construct from the current entry level right up to the apex.

3. **Macrososi**: The vibrational trader executes upsells as price rises across one or more intervals, followed by downbuys if price should decline. This method is bounded as long as it takes place below the apex. Macrososi cannot exit above the apex as there will be no more shares or positions left to scale out of. The upsells must not violate the Upsell Rule—the volume of upsells must not exceed the volume of all the downbuys—otherwise the trader would be effectively establishing a net short position, which is not allowed under the principles of directional boundedness.

Implications of Upside Microsiso on Trend Capture

The Upside Microsiso is a mechanism whereby upbuys are followed by higher upsells across a single interval. Within that mechanism, you cannot capture trending action if you exit the long position. From the principle of the null positional, we learned that a buy-sell at the same price level equates to no net change and is the same as holding the buy all the way up. As such, there is no point in using the Upside Microsiso mechanism to catch trending action. A simple buy and hold will suffice! See Figure 7.5 below.

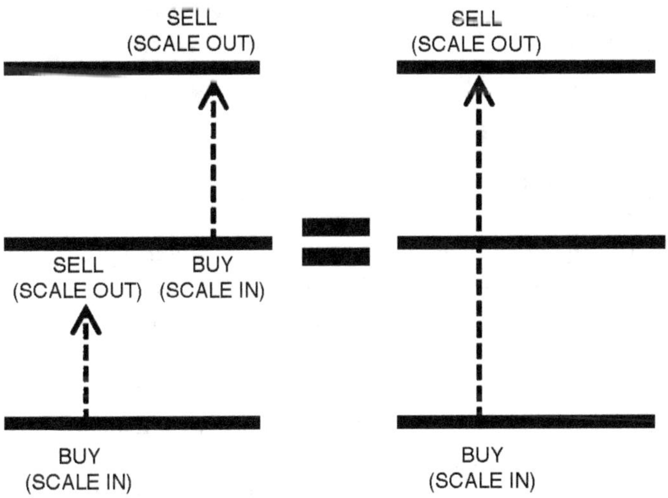

FIGURE 7.5 Redundancy of the Upside Microsiso

Scaling Mechanisms With Reference To Entry Position

Scaling across multiple intervals to the downside is considered unbounded if the total allocated capital is unaccounted for or unknown. Scaling in across multiple intervals to the upside is also unbounded if the total allocated capital is not accounted for in advance. As we can see from Figure 7.6, upside scaling in of new longs is unbounded unless the maximum number of long entries is known.

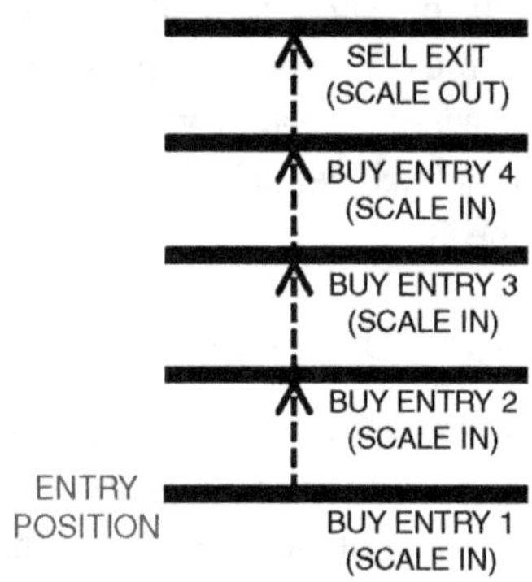

FIGURE 7.6 Unbounded Macrosiso Trading

The Macrosimo and the Unbounded Upside Macrosiso mechanisms violate the principle of order entry boundedness by using stop orders for entries and exits.

RANGE BOUNDEDNESS

Trading with respect to relative range extremes rather than absolute price extremes will usually result in erosion in capital, as stop losses will deplete capital every time they are hit. By not pegging the relative range trading extremes to absolute price extremes like zero test level, unbounded traders have a high probability of getting their stops hit. This is because the stop loss levels are in a sense only relative, or floating, within the historical price range of the stock or commodity.

Capital boundedness, which is all about the preservation of capital, requires that a vibrational trader have sufficient disposable capital to sustain near zero test level. This total allocated capital must be available and dispensable at all times. Since the lower vibrational trading range is an absolute price extreme, there may also be a natural support present, be it in stocks or commodities. This is because there is a general agreement as to the price of this extreme, as opposed to the relative insignificance of an arbitrary trading range that is not pegged to the absolute extremes. That arbitrary range is unbounded, which causes trading capital to be unbounded as well.

Relative Trading Range vs. Absolute Price Extremes

To preserve capital, or keep it in a bounded state, you must trade price extremes with sufficient disposable capital. The reason we trade the absolute extremes in price is because it give us a reference point in price as to the maximum allocated capital required to sustain a zero test level event. If you can afford to trade to the zero test level, it will only be bounded if your capital can work in the markets indefinitely. If your disposable capital is insufficient to allow trading to the zero test level, that will result in unbounded trading as well. This is also the case if the trader did not have enough time to replenish or dispense capital, resulting in profit driver exits before price extreme is met.

Implications of Range Boundedness

Refer to Figure 7.7.

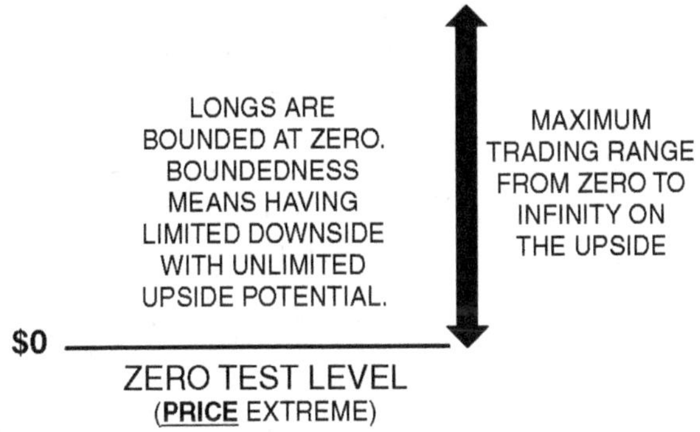

FIGURE 7.7 Price Extreme in Range Boundedness

- Losses for long orders, including long options, are bounded at the zero test level since full face value has already been paid up.
- Stops are not used in bounded trading, therefore there are no losses from oscillation or gapping.
- There is a tolerable amount of capital lost in the unlikely event of zero test, since the capital used is disposable.
- You can extract full profits without closing all long positions, as profits are measured as a change in market value and not as running profit and loss.
- There is a low probability of price reaching zero compared to the chances of a stop being hit in an arbitrary price range.
- The use of limit orders only increases trading performance.

Implications of Trading an Arbitrary Range Extreme (Unbounded)

Refer to Figure 7.8.

FIGURE 7.8 Arbitrary Range Extreme in Unbounded Trading

- The use of stop loss orders to avoid sustaining intolerable losses in turn introduces capital bleed.

- There exists an intolerable upper and lower boundary for shorts and longs, between which there are oscillation losses.
- The possibility of losing all capital is high with consistent equity bleed at every stop loss exit.
- You are unable to extract full profits without closing all positions in margin accounts.
- There is a higher probability of price hitting an arbitrary trading range extreme than a price extreme.
- Use of stop orders degrades trading performance as a direct result of negative slippage.

Extending and pegging the relative trading range to zero test level means that there should be sufficient capital to hold positions down to that price extreme. Since, theoretically, we cannot accomplish this with shorts on the upside, the vibrational methodology will only initiate long positions . Therefore, range boundedness also lends itself to directional boundedness. In effect, we have therefore bounded our risk capital at zero test point by pegging it to the lower price extreme.

ORDER ENTRY BOUNDEDNESS

The Principle of Order Entry Boundedness requires that entry and exit orders be used in a way that does not lead to capital depletion. This immediately precludes the use of all forms of market orders, including stop entry and stop exit orders. Market orders are considered unbounded mainly due to the fact that these orders are adversely affected by price gapping action. A buy or sell stop cannot protect the market participant from losses. Similarly, stop entry orders cannot prevent negative slippage, which directly affects the conservation of total allocated capital. Hence, only limit entry and exit orders are allowed under the principles of capital boundedness.

For Bounded Entries

A buy limit order allows you to buy at a predetermined price or better (lower). This is excellent for trading the SISO/SOSI constructs as you can only be exposed to positive slippage of limit orders, giving higher profits and higher PLR, BLR, and MLR. Bounded entries also will give the vibrational trader the edge in entering the pyramidal SISO/SOSI levels, especially if price overshoots and gives the trader an even lower entry price.

Hence, we welcome extreme volatility when using limit orders as we can only do better. Should the order not be filled, then we let time work for us until it fills at another SISO/SOSI level. Price will have to traverse to one of the levels eventually.

For Bounded Exits

A sell limit order allows you to sell at a predetermined price or better (higher). This is excellent for trading the SISO/SOSI constructs as you will always be exposed to only higher profits. We welcome extreme volatility when using sell limit orders since, again, we can only do better. Should the order not be filled, then we let time work for us. There is no escape for price. It will have to traverse one of the levels eventually and until then we use scalping techniques to extract interim profits.

In short, a buy limit will give you a better entry level from which to execute your income strategy, especially if price overshoots and gives a cheaper fill. By the same token, a sell limit will give you a better exit level if price overshoots and gives you a better fill.

> *We position our entries and exits advantageously with buy and sell limit orders.*

Limitations in Boundedness

- In range boundedness, the price exposure is limited by the zero test level.
- In capital boundedness, maximum potential capital exposure is limited at paid-up full market value.
- In market timed boundedness, the timing of market behavior is limited only to volatility.
- In directional boundedness, trade direction is limited to long exposure and no naked shorts or net short positions are allowed.
- In order entry boundedness, the maximum price paid for entry is also limited via the use of limit orders.

This essence of "limiting" is the very cornerstone of the boundedness trading philosophy.

Finally, let us summarize the nature of boundedness and its function:

- Range boundedness ensures that our trading capital works over the entire price range, regardless of market conditions.
- Capital boundedness ensures that our capital stays intact (and not depleted) so that we keep profiting continually in volatile and unpredictable markets. The more time invested, the more income is generated.
- Market timed boundedness ensures that capital works efficiently by exposing the vibrational constructs to more volatile market action.
- Directional boundedness ensures that we always trade under the rules of range boundedness as we cannot limit upside short losses without capital bleed.
- Order entry boundedness ensures that our entries and exits are as efficient as possible in maximizing profits and limiting losses.

We have therefore set up the ideal conditions for both time and money to work for the benefit of the vibrational trader, and not against.

Unboundedness in Vibratrading

We shall be using bounded as well as unbounded mechanisms in vibratrading. But a distinction must be made here as to the type of unboundedness used.

The only type of unboundedness in vibratrading is Order Entry Unboundedness. We employ buy and sell stop entries and exits in many strategies, but we will never use a stop loss order to exit in loss.

We may only use stop orders to:

1. Lock in a profit upon exit
2. Achieve a breakeven entry
3. Enter the market.

Running and Working Capital at Zero Test Level

As you know from the equity valuation analysis in the preceding sections of this book, maximum loss in bounded trading occurs when all the longs hit or test zero test level. This can only happen if the share price of all the component stocks in the fund's basket decline to a value of zero. The trader cannot lose more than the total allocated capital.

At the zero test level,

$$\begin{aligned} \text{Total Loss} &= \text{Full IMV of all Long Positions} \\ &= \text{Total Allocated Capital.} \end{aligned}$$

$$\text{IMV} - \text{CMV} = \text{Maximum Running Capital} = \text{Maximum Working Capital.}$$

Running capital is the difference between the IMV and CMV of all longs in a vibrational construct, while working capital is the IMV of open longs in a vibrational construct.

CHAPTER 8

The Mechanics and Dynamics of Vibratrading

In the following pages, you will learn about the various bounded and unbounded vibrational and trend mechanisms. But I feel it is extremely important to first put everything into perspective by showing you where and how the mechanisms fit into the larger vibrational constructs. After getting acquainted with the overall purpose of the various mechanisms, we may then delve into the workings and construction of each of them.

The objective of building a vibrational construct is to extract oscillational as well as directional profits from the markets. Though the directional mechanisms are not truly bounded, we still introduce some of these techniques so that the vibrational trader may profit from trending markets as well.

Having said that, we strive for the ultimate profit extraction methodology, one that profits whether the market rises, falls, or flatlines. This non-directionality arises from two mechanisms:

1. Collection of options premiums from the selling of Covered Calls and Naked Short Put Options within the vibrational construct,
2. The oscillation profits generated from the various Micro and Macro SISO/SOSI mechanisms.

VIBRATIONAL OPERATIONS, MECHANISMS, AND CONSTRUCTS

Vibrational mechanisms cover the most basic elements of profit extraction via price oscillations. They are comprised of simple entry and exit scaling operators. There are eight basic scaling operations—long and short:

- Downbuys
- Upsells
- Upbuys
- Downsells.

Long upbuys and downsells are singled out as unbounded operators and belong to the unbounded Micro/Macrosimo trend capture mechanism. The combination of these eight operators gives rise to what is called a Vibrational Construct.

Vibrational Constructs

Below is a summary of all seven bounded and unbounded vibrational and trend constructs and their corresponding scaling operators, for your easy reference.

1. **Bounded Microsiso:** Downbuys are initiated as price falls and are immediately followed by a corresponding upsell across only one interval. This mechanism can be initiated anywhere within the pyramidal structure. It is also found above the capstone entry level after a full Macrososi swing from the apex, as well as in a bounded upside Macrosiso should price reverse to the downside. Only limit orders are used for entries and exits.
2. **Bounded Macrosiso:** Identical to the bounded Microsiso except that subsequent upsells are across two or more intervals. Only limit orders are used for entries and exits.
3. **Bounded Upside Macrosiso:** A downbuy is initiated during an uptrend on price throwbacks to the entry levels, and is followed by an upsell across two or more intervals. It is fully bounded because the total allocated capital for entries up to the apex has already been accounted for in the methodology. As a result, this mechanism is a form of bounded trend and vibrational capture. Only limit orders are used for entries and exits.

4. **Bounded Upside Microsiso:** This mechanism is similar to the bounded upside macrosiso, the only difference being that the upsell takes place across single intervals. Again, only limit orders are used for entries and exits.
5. **Bounded Macrososi:** An initial large buy or downbuy is followed by smaller upsells across every one interval. The mechanism essentially scales out long positions at successively higher levels toward the apex, where it terminates. It may be initiated anywhere within the pyramidal structure and may extend all the way up to the apex.
6. **Unbounded Upside Macrosiso:** Upbuys are followed by higher upsells across two or more intervals. It is unbounded, as buy stop and sell stop orders are used to enter and exit positions. The total allocated capital for all intended upbuys must be accounted for within the design of the construct.
7. **Unbounded Macrosimo:** The upbuys and subsequent higher upsells are executed across two or more intervals. This mechanism is unbounded as it violates the Principle of Order Entry Boundedness by using market orders to enter and exit the markets.

These are the seven mechanisms that populate the vibratrader's toolbox. But we shall not be focusing on these seven mechanisms alone. There are numerous other trend and vibrational constructs in the chapter entitled "Putting It All Together."

Vibrational constructs are a function of combining the eight operators to achieve an overall profit objective. The constructs are put together with regard to specific parameters which include:

1. Sharesizing
2. Scale factoring
3. Capstone share multiples
4. Interval sizing
5. Retracement target objective
6. Total allocated capital
7. Effective range of the instrument traded.

The vibrational mechanisms will adopt the overall parameterization and functionality of the factors listed above. A vibratrader can use the constructs to control the speed, magnitude, and bias of the expected returns and profit. The speed and magnitude of the returns will depend largely on the total allocated capital available to the vibrational trader or investor. Some vibrational constructs are more capital intensive than others. Three of the constructs discussed in this book are:

- Null-based
- Profit-based
- Phi- or Martingale-based.

The idea of a Martingale-style vibrational construct is best seen in the Phi-based vibrational construct. Each successively lower downbuy level deploys a certain amount of shares that will completely cover all running capital within a single vibrational upsell. The implementation of this type of doubling-up or cost-clearing strategy is only possible on instruments that are unlikely to test zero, like equity ETFs, stock market indices, and commodities. A vibrational trader can finally ascertain the maximum sharesize required near the zero test level, and this will be at or near the Martingale limit. In essence, we have a bounded Martingale strategy.

The Significance of the Capstone Entry

The pyramidal structure is the final result of all vibrational operations, mechanisms, and constructs (see Figure 8.1).

The structure takes on a pyramidal appearance due to the application of boundedness which dictates that an upsell/downbuy mechanism be used. As price falls, we build more share exposure via the downbuying. As price rises, we exit shares via the upselling. This creates a pyramidal structure representing our exposure in the market.

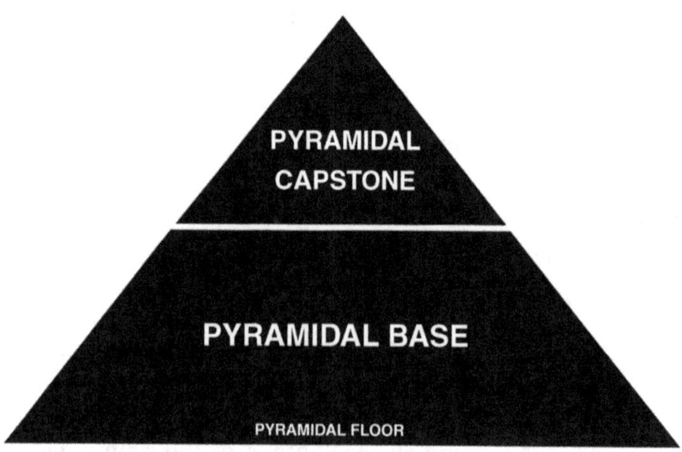

FIGURE 8.1 The Geometry of a Bounded Vibrational Structure

The capstone entry level is the initial vibrational entry into the pyramidal structure. As such, the capstone entry level represents the reference by which the total allocated capital is calculated for the specific vibrational construct required. The capital must be sufficient to fund long positions across the entire pyramidal structure, ending at the apex. No additional capital is required. The capstone sharesize will determine the sharesize at all other levels above and below it, via the scale factor.

THE SCALE FACTOR

The scale factor determines whether a vibrational construct is more biased toward trend or vibration. It is probably the most important parameter in the construction of a vibrational structure. The profit potential of a construct is directly related and proportional to the capstone sharesize, scale factor, and interval size. The higher the scale factor, the greater the returns as price approaches the lower levels of the construct.

The width of the sharesize diagram represents the number of shares at each level. The width of the pyramidal construct is a function of the capstone sharesize and the scale factor. The wider the sharesize diagram, the greater the profit potential, but also the greater total allocated capital.

The height of the sharesize diagram indicates the depth or range of the traded ETF or stock. Higher PLR and BLR values are usually associated with shorter sharesize diagrams.

Some example sharesize diagrams (see Figure 8.2) may help illustrate this concept.

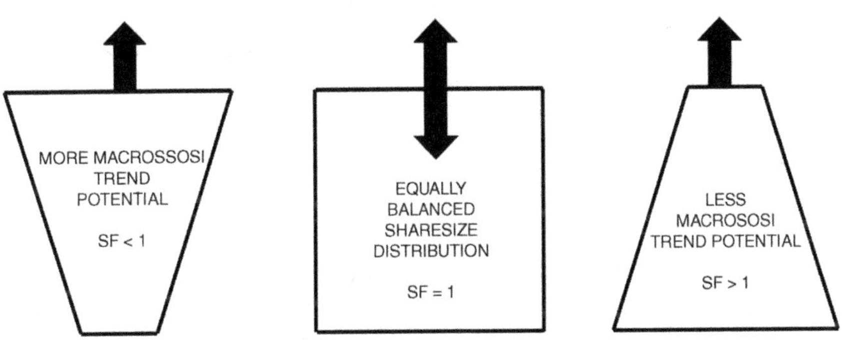

FIGURE 8.2 Sharesize and Trend

Scale Factor Stability

Every structure has a point representing its level of stability in a system. It is usually referred to as the center of gravity. Our vibrational constructs are no exception. The "center of gravity" of our structure is represented by the average price. All things being equal, the lower the average price, the greater the vibrational returns. A low price forms a wide and very stable base, as shown in Figures 8.3 and 8.4 below.

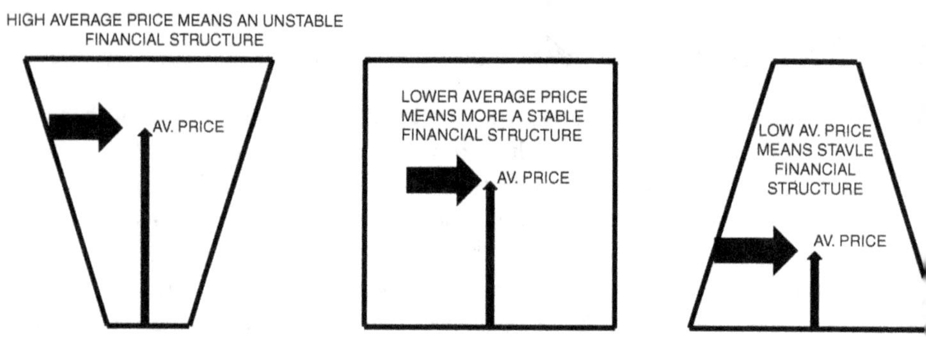

FIGURE 8.3 Sharesize and Stability

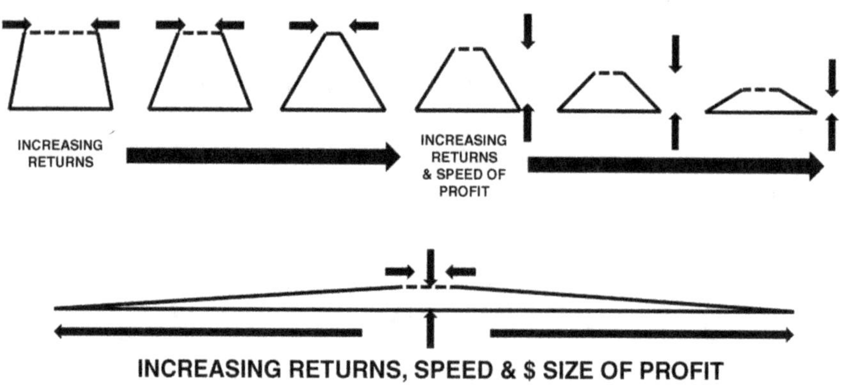

FIGURE 8.4 Sharesize and Return Characteristics

Interval Sizing

Extracting profit from market volatility, the crux of vibratrading, means profiting from price oscillations. Profiting from price oscillations means getting price to trigger the upsells and downbuys as frequently as possible.

In order to achieve this, we can select a more volatile market. Alternatively, we can reduce the interval size by adjusting the scaling sensitivity to price action. In Microsiso and Micrososi trades, every trade builds capital since every exit is profitable as a consequence of boundedness. However, not every price motion triggers a trade. The size of the interval plays a large role in achieving consistency in vibrational returns.

When price motion triggers an entry, which is always a downbuy in SOSI/SISO trades, we say that we have a position. When a downbuy exits in an upsell, we say that we have a trade, and in the process we monetize profit. Our objective is to trigger more trades. Unfortunately, "slip through" reduces the number of trades triggered due to either trending or sideways action lacking large enough oscillations in price.

Maximizing Vibrational Profits and Scale Sensitivity

Here are some general guidelines relating to scaling sensitivity.

Generally, for wider intervals in the construct, we expect:

- Lower capital requirements due to the smaller number of downbuy levels.
- Lower scale sensitivity as price is required to traverse larger distances in order to trigger downbuys and upsells.
- Higher break-even price levels since there is less averaging down.
- Lower returns due the decreased scale sensitivity.

With a higher scale factor, we expect:

- Higher capital requirements due to the greater number of shares purchased as price declines.
- No change in scale sensitivity assuming the interval size remains constant.
- Lower break-even price levels since there is more price averaging with the larger number of shares bought at lower levels.
- Higher returns due to the increased sharesizes.

With more shares traded in the construct, we expect:

- Higher capital requirements.
- No change in scale sensitivity assuming the interval size remains constant.
- Change in the break-even price levels if the scale factor is not one.
- Higher profit due to the increased sharesizes.

CAPSTONE MECHANISMS

All trade entries must take into account the possibility that price will rise or fall; price action is unpredictable. Therefore, all trade entries must be ready to assume different scaling parameters to be accountable for the total allocated capital required in a zero test level event. This is a necessary condition for boundedness. Hence the Initial Market Value (IMV) must represent the maximum initial capital investment allowable at the capstone entry level, which forms the start for every Macrosiso, Microsiso, and Macrososi trade. That level represents the base of the smaller pyramidal scaling construct at the top end of the pyramidal structure and is the initial entry level for all bounded trades (see Figure 8.5). If all the principles of boundedness are preserved, we then say that the IMV is capstoned.

Capstone-based constructs are both vibrational and directional in nature.

The two scaling mechanisms are:

1. The Scale Out–Scale In (SISO) vibrational mechanism called Macrososi.
2. The Scale In–Market Out (SIMO) directional component called Macrosimo.

Note that the vibrational method above is bounded while Macrosimo breaches the principles of order entry boundedness.

Let us now turn our attention to the first vibrational mechanism, the Macrososi mechanism.

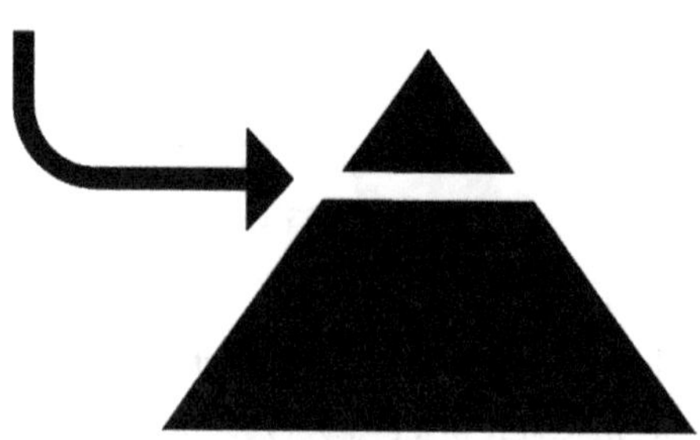

FIGURE 8.5 The Capstone Entry Level

THE MACROSOSI VIBRATIONAL MECHANISM

The Macrososi mechanism captures both trend and vibrational components of price action. It is primarily a scale-out mechanism. A trader starts a number of long positions with the maximum capstone IMV allowable, and then proceeds to scale out as price rises from the capstone entry level. Upon rising to the first upsell level, the mechanism may now begin its subsequent downbuys, re-entering the long position. Interestingly, the Macrososi mechanism is usually a temporary construct as it transforms into a Microsiso or Macrosiso mechanism as soon as price hits the apex of the pyramidal construct and begins to descend.

Please note that the sharesize diagram seen in Figure 8.6 is a geometrical representation of the Macrososi mechanism. We see the sharesize diagram tapering off at the top as it approaches the apex, due to the fact that the number of shares is gradually diminishing from scale outs.

Generally, Macrososi constructs posseses a bottom-heavy sharesize structure since the maximum sharesize is at the capstone entry level and shares begin to scale out as price rises. The initial sharesize at the capstone represents the maximum shares tradable for the total allocated capital assigned. As price oscillates within the capstone, vibrational profits will

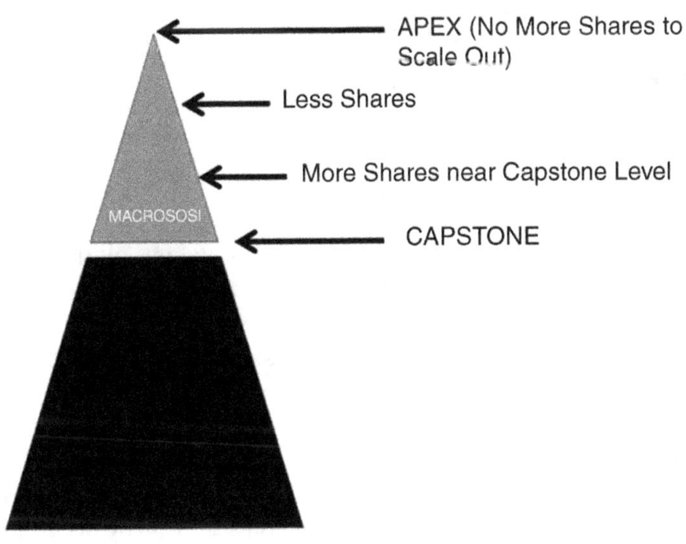

FIGURE 8.6 Macrososi Sharesize Diagram

be continually extracted from the markets. This vibrational trading range extends from the capstone entry level right up to the maximum trading price level for Micro/Macrososi trades, which happens to be at the apex of the capstone. Remember, upsell means "selling to exit" an old downbuy position as price rises above, and a downbuy "buys to open" a new position as price falls to the next entry level (see Figure 8.7). We sell on the way up and only start to buy if price declines to previous lower levels.

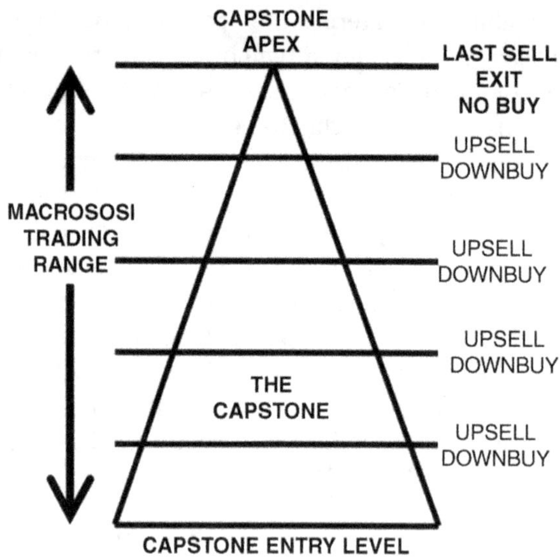

FIGURE 8.7 Macrososi Construct

Macrososi Profit Extraction

Macrososi upsells all the way to the apex and downbuys all the way down to the capstone entry level. If price penetrates the capstone entry level to the downside, it will turn into a Microsiso or Macrosiso trade. A full cycle of upsells and downbuys from the capstone base to the apex and back to the base again is called a full capstone swing. Assume all intervals are equidistant, for convenience. The minimum profit occurs when price goes directly to the first upsell and directly back to the capstone base in a single oscillation. The maximum profit, on the other hand, is indeterminate, as it occurs whenever there are upsells and downbuys caused by price vibration within this capstone construct.

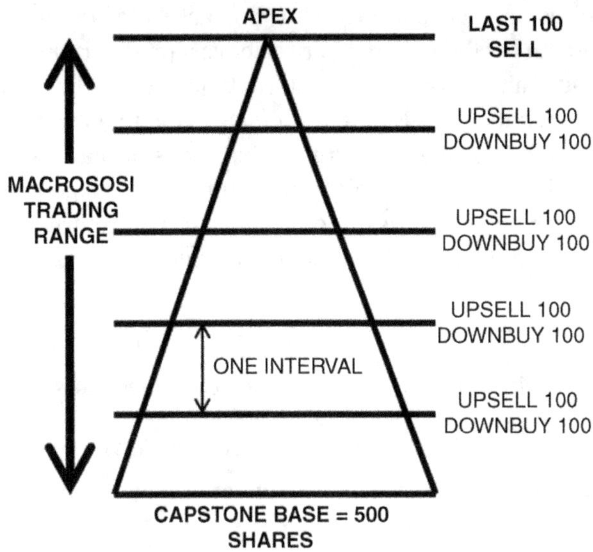

FIGURE 8.8 Example Macrososi Construct

> *Hence, we require volatility or price vibration in order to continually extract or generate returns from the markets.*

Using Figure 8.8 as a reference, let's assume that price triggers the very first upsell level and that each price interval is $1. With an initial capstone sharesize of 500 shares, the vibrational trader sells the first 100 shares. This equates to a return of $100, since the return is equal to the number of shares multiplied by the interval value $1, giving us a $100 return. The next upsell gives the trader $100 shares at a $2 interval, returning $200. Therefore, the total return for a direct price run up to the apex will be $100 + $200 + $300 + $400 + $500 = $1,500.

Now, if price declines from the apex, the vibrational trader may begin to initiate downbuys at the same levels or lower than the original upsells. Assuming that the trader buys into the exact same levels on the way down, a total of $400 + $300 + $200 + $100, or $1,000, will have been reinvested in the new Micrososi construct. This leaves the vibrational trader with a full Macrososi swing profit of $1,500 – $1,000, or $500.

Note that as long as price oscillates between the capstone and the apex, vibrational profits will be continually extracted from the market. The vibratrader does not need the market to rally or fall in order to generate returns, in stark contrast with "buy and hold" investors who wait with bated breath for their shares to rise above a predetermined break-even level or average value in order to profit. No time is wasted waiting for these uncertain eventualities in vibrational trading. Vibrational returns are continually generated regardless of future price levels or market action of the stock or ETF. Note also that no further returns may be generated should price extend beyond the apex. To extract further returns, we use the vibradirectional strategies outlined in a subsequent chapter.

Macrososi Average Price Drift

Average Price Drift may occur in the Macrososi mechanism if:

1. The upsell levels differ from the subsequent downbuy levels, or
2. There is a difference between the numbers of shares bought and sold at the same upsell and downbuy levels.

Refer to price path A in Figure 8.9.

Once the price path has completed its full swing, the average price of the structure will drift upwards. Note that before any upsells were triggered, we had 500 shares at an average price of $1. Upon completing its full swing, the vibrational trader now owns 100 shares at every price level from $1 to $5.

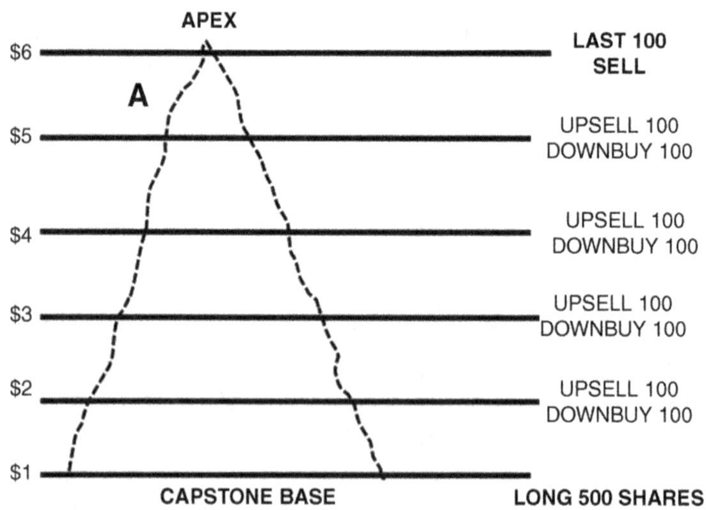

FIGURE 8.9 Macrososi Average Price Drift

Hence the average price is equal to ($5 + $4 + $3 + $2 + $1) × 100 shares/500 shares = $3.

The average price drifted from $1 to $3 upon one full swing.

Macrososi Full Swing Consequences

The trader started off with an IMV of 500 × $1 = $500. After one full swing, the CMV is now ($5 + $4 + $3 + $2 + $1) × 100 shares per level, or $1500. Therefore, even though the trader made $500 in the full swing, an additional $1000 in capital was subsequently required to maintain downbuy positions and, more importantly, the capstone's vibrational construct. Note also that if price made three full oscillations before returning to the original $1 price level, the trader would have made $500 × 3 swings or $1,500, thereby reducing the cost basis of all the shares held within the vibrational capstone construct to zero. The vibrational trader now has a risk free mechanism with which to generate vibrational profits within the capstone indefinitely.

This rise in average price drift must be accounted for at the initiation of the Micrososi trade. It is a condition of sufficiency required by capital boundedness, even more so if scale level and size differ in the Macrosiso on the way down from the apex.

Remember, the Macrososi mechanism will cease to generate any vibrational returns beyond the apex, as there will be no additional shares to upsell. If price starts to decline from the apex, the Macrososi mechanism will be permanently transformed into a Microsiso or Macrosiso mechanism, as seen in Figure 8.10.

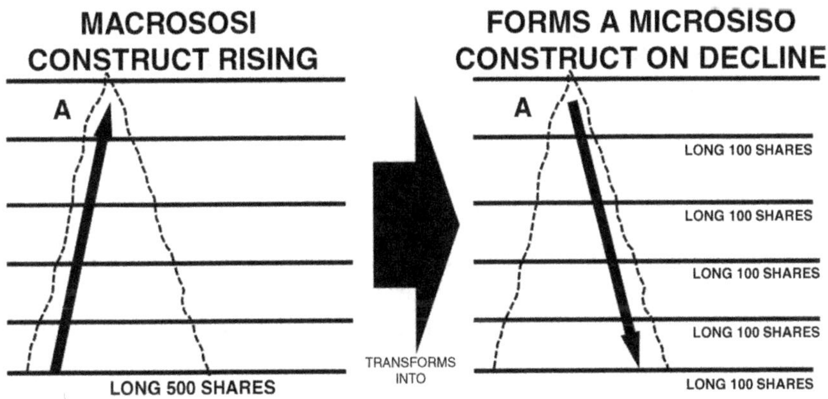

FIGURE 8.10 Macrososi Full Swing

MACROSIMO MECHANISM (UPBUY - UPSELL)

Simply, "simo" stands for Scale In-Market Order Out. The Macrosimo mechanism possesses a top-heavy structure since its maximum sharesize is at the highest position, possibly even beyond the apex. It begins to scale in upbuys as prices rise. Its entry also represents the start of a potential Bounded Microsiso or Macrosiso trade into the main pyramidal body. All Macrosimo trades may begin at either the capstone entry level or the apex.

Being primarily an upbuy–upsell mechanism, the upsells in Macrosimo are usually withheld over two or more intervals. As there is no buying limit to the upside, it is capital unbounded and we cannot ascertain the required capital to keep buying indefinitely! This problem is somewhat resolved over the next few pages, but the mechanism still fails to be strictly bounded, even as a capstone-based entry, since order entry boundedness has been violated. However, we deal with this directionally via the use of the Self-Financing Share Purchase (SFSP) approach in order to alleviate the need for more capital to finance an essentially undefined upper buying limit.

There are three approaches to capturing trend profits via the Unbounded Macrosimo mechanism, namely:

1. The Profit Trailing Exit
2. The Monetized Risk Exit
3. The Discrete Exit Method.

All three methods may use the SFSP technique to gain exposure in the equity markets with little capital.

Self-Financing Share Purchase (SFSP) Mechanism

SFSP is designed to facilitate the self-financing of subsequent share purchases via the securitization of excess available funds in a stock margin account. As such, the investor will require only minimal "starter" capital to account for the trade's very first margin requirement and capstone IMV. This system will only halt when the trader lacks sufficient buying power or all the available funds are utilized.

SFSP is considered an unbounded strategy because the total allocated capital required for its construct is unknown. It can be converted into a safer form of trading via the Profit Trailing Exit technique. As long as average price is not hit, the share purchases will be continually and indefinitely self-sustaining.

SFSP is also an important concept for odd sharesize and partial sharesize trading, where sharesizes can be very small, even as low as a few shares in many cases. It instructs the trader on the required minimum purchase of new shares that are self-sustaining, using the account's unrealized or unmonetized profits. The trader just needs to place the starter sharesize and the mechanism will self-finance additional purchases. It is in effect buying on excess margin (see Figure 8.11).

In the example below, we see the self-financing mechanism in action. Assume that the trader wanted to self-finance one new share for every \$1 price increment. As seen from the table above, the trader is required to buy 11 shares at \$10 per share. When the share price reaches \$11, the trader will have an extra \$11 to purchase or finance 1 new share for a purchase price of \$11. The trader now has 12 shares at \$11 per share. Once the price reaches \$12, the trader will have an extra \$12 to purchase a single extra share at \$12. This process will continue as long as price continues to rise.

The only issue with this exercise is that if price declines, the trader will be holding on to large losses due to the new positions falling below their respective entry levels. To avert this, the trader must implement

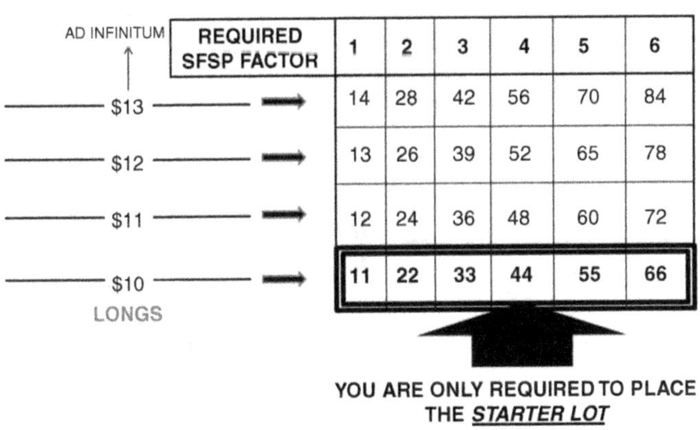

FIGURE 8.11 Self-Financing Share Purchase
Note: SFSP Factor is based here on \$1 increments

some kind of downsell operator, in this case the ubiquitous stop loss mechanism.

So, how do we calculate the sharesize required to initiate a self-financing mechanism?

First, decide on the SFSP Factor required. (This means the amount of shares you wish to purchase at each higher price level.)

Starter Sharesize = (Entry Price × SFSP Factor) + SFSP Factor.

Example: Entry Price = $10.

Required Factor = 5 additional shares for every $1 rise in price.

Therefore, starter sharesize = ($10 × 5) + 5 = 55 shares.

Referring to Figure 8.11, we see column 5 indeed indicates a minimum purchase of 55 shares. From then on, the system will self-finance all additional shares at the rate of 5 new shares per every $1 increase in share price.

Try this for an SFSP exercise: Fill up the table in Figure 8.12 with the correct sharesizes for the required SFSP Factors. (Try not to look at the answers in Figure 8.13 first.)

Answers

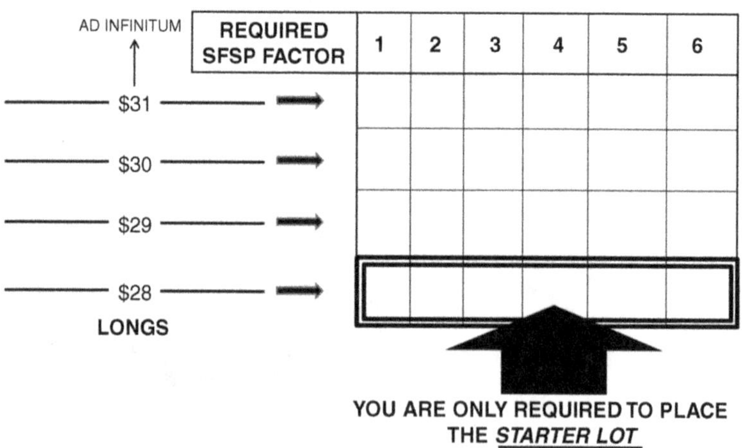

FIGURE 8.12 SFSP Exercise

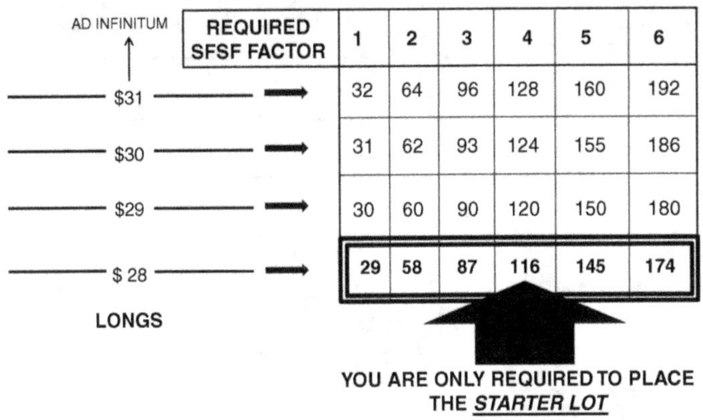

FIGURE 8.13 SFSP Answer

Profit Trailing Exit

To make SFSP work more safely, we need to establish the average Break-even Level in order to exit beyond the average price to achieve a profitable exit. The calculations for average price are again given in the following equation:

$$\text{Average Price} = \frac{\text{CMV}}{\text{Total Shares}}$$

In Figure 8.14, we see that in order to self-finance 10 new shares for each \$1 rise in price, we are required to purchase 110 shares at \$10 each. Upon reaching \$11, the trader would have \$110 to purchase 10 new shares at \$11, and so on. The average price formed between the initial entry at \$10 and the new entry at \$11 is simply (\$110 + \$1,100)/120 = \$10.08.

As price rises, the average price also begins to rise, but at a slower rate than price itself.

As long as the sell stop exit orders are at or beyond the average price, and we discount gapping and slippage, the SFSP is essentially risk free. We shift our exit point above the average price to lock in a risk-free profit. We keep trailing price until the upsell is hit. This is the unbounded trend mechanism in action, executing upbuys with minimal capital, and exiting positions above the average price via the use of a trailing sell stop. Hence the name: Scale In (Upbuys) – Market Out (Upsells).

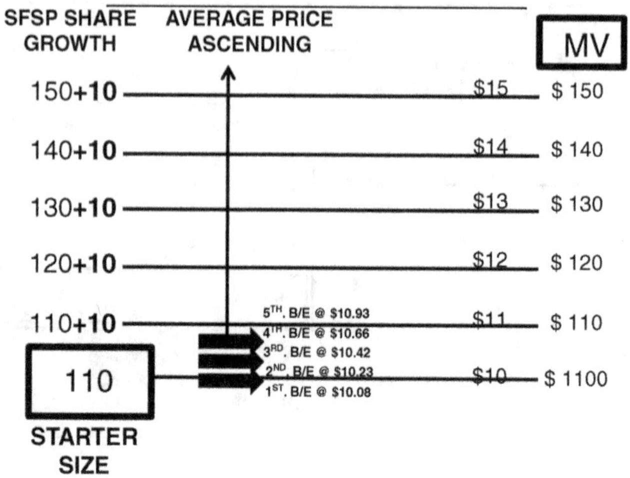

FIGURE 8.14 Profit Trailing Exit

Monetized Risk Exit: The Macrosiro Mechanism

In the profit trailing exit, we liquidated all our shares at profit. But we could also just monetize the risk, giving us a breakeven exit with risk-free shares working in the market. The monetized risk exit, or Scale In-Risk Out, is extremely simple. You start off with a Macrosimo trade, but instead of arranging to be stopped out, you monetize risk at an arbitrary level set anywhere above the average price. Use the formula below to determine how many shares to liquidate.

Shares to Liquidate = Total CMV/Exit Price

Macrosimo Discrete Exit

The third way to exit the Macrosimo is to exit at one or more upsells above the corresponding upbuys. In the profit trailing exit, we liquidated all our shares at profit. In the Discrete Exit, long positions are exited one at a time. Each position has locked-in profits at the next upbuy using a sell stop order. The trade starts as a Macrosimo profit trailing exit and converts into a discrete exit (see Figure 8.15).

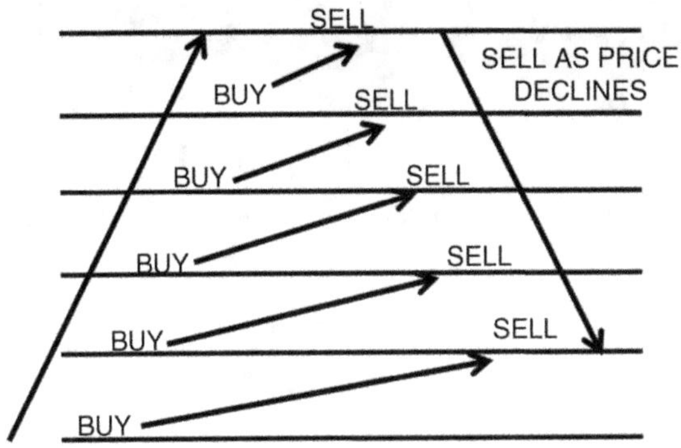

FIGURE 8.15 Macrosimo Discrete Exit

CHAPTER 9

Pyramidal-Based Vibrational Mechanisms

In the last chapter we covered capstone vibrational mechanisms. There are four bounded vibrational mechanisms within the pyramidal base, namely:

1. The Microsiso
2. The Macrosiso
3. The Upside Macrosiso
4. The Upside Microsiso.

Let us examine the first Pyramidal-based mechanism.

MICROSISO

The Microsiso possesses a bottom-heavy structure since its minimum sharesize is at the capstone entry level, and it begins to scale in more shares as prices fall. All Microsiso trades begin either at the capstone level or from the apex, especially after a Macrososi full swing. It also represents the start of a potential Macrososi or Macrosiso trade. Note that

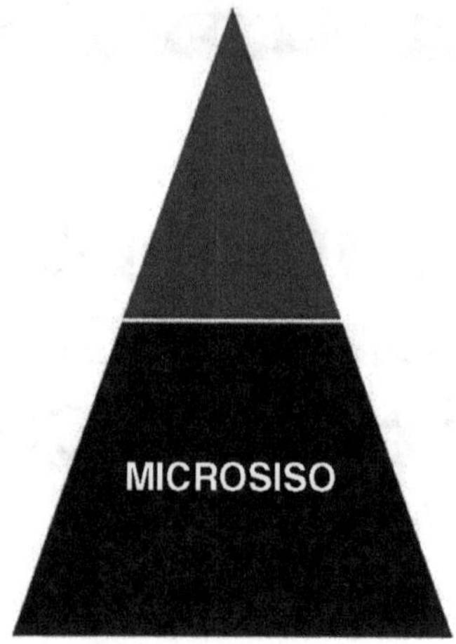

FIGURE 9.1 Microsiso Within the Pyramidal Construct

should price oscillate within the pyramidal base, vibrational profits will be continually extracted from the market via the Microsiso mechanism. The Microsiso extends from the apex or capstone entry level right down toward the floor of the pyramidal structure (see Figure 9.1).

As seen in Figure 9.2 below, the Microsiso works by executing downbuys and upsells across single intervals only.

We shall now turn our attention to share and profit accumulation using the SISO Mechanisms.

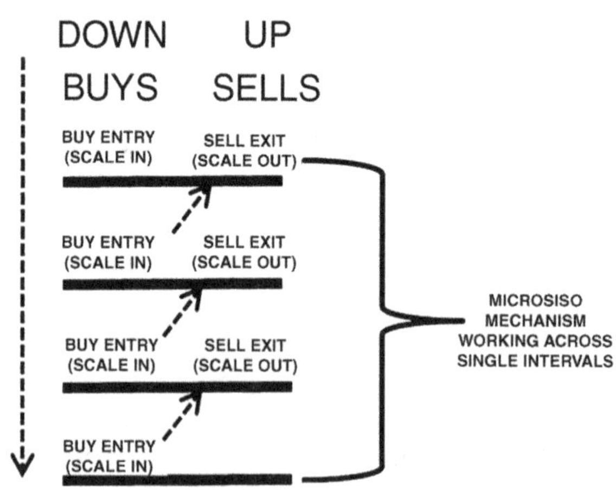

FIGURE 9.2 The Microsiso Mechanism

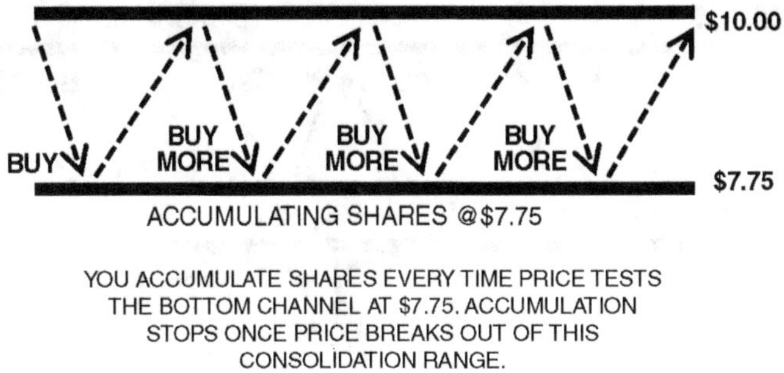

FIGURE 9.3 Share Accumulation During a Consolidation

SISO-Based Share Accumulation

Let us assume that you would like to accumulate a certain stock during its consolidation phase, before it breaks to the upside, as in Figure 9.3 above.

Every time price hits $7.75 you buy a little more, realizing as you do that your current value at risk, or CVAR, is increasing. Finally, you have 40 shares when the stock plummets. You immediately sell and get out—you suffered a long squeeze. That running capital is lost permanently. This scenario occurs over and over, depleting all your capital in the process.

Now, let's do this in another way (refer to Figure 9.4).

Assume that you invested $310 in 40 shares with an average price of $7.75 in one transaction.

$$\begin{aligned} \text{IMV} &= \text{Average price} \times \text{Total shares} \\ &= \$7.75 \times 40 \\ &= \$310 \end{aligned}$$

The current price is now at $10. You proceed to monetize IMV by dividing by current price ($10), which means you sell 31 shares with 9 left in

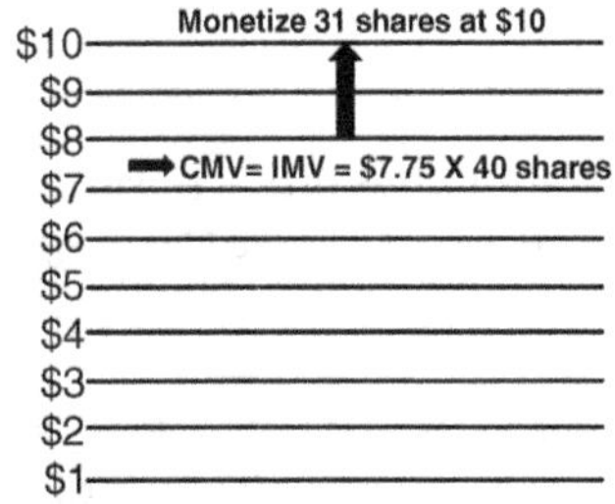

FIGURE 9.4 Monetizing IMV

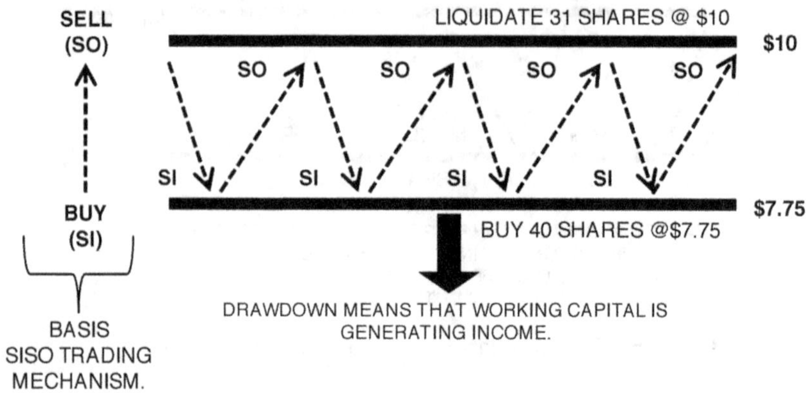

FIGURE 9.5 Risk-Free Share Accumulation

the market. Price then declines to the original average price of $7.75 and you now securitize $310 at $7.75 and buy back the 40 shares. You have just accumulated nine additional shares for free! You may leave them to appreciate in value or repeat the process and keep accumulating more shares. Let the market vibrations help you accumulate shares risk free, alleviating the need to worry about future price direction. If the stock plummets, this trade becomes part of a capstone-based entry and you start your pyramidal Micro/Macrosiso trading. No capital bleeding occurred anywhere during the entire process (see Figure 9.5).

We compare the two accumulation approaches side by side in Figure 9.6.

So, instead of buying shares with your own capital at each oscillation, you collect free shares at each oscillation. The only reason this can work without potential capital depletion is because it is part of a capstone-based entry into the pyramidal SISO/SOSI construct.

SISO-Based Profit-Locking

Say you want to accumulate some profit during the consolidation phase before it breaks out. Every time price hits $10, you monetize the profit. The number of shares to liquidate is easily calculated as follows:

$$\begin{aligned}\text{Profit/Current Price} &= \text{CMV} - \text{IMV/Current Price}\\ &= (\$400 - \$310)/\$10\\ &= 9 \text{ shares.}\end{aligned}$$

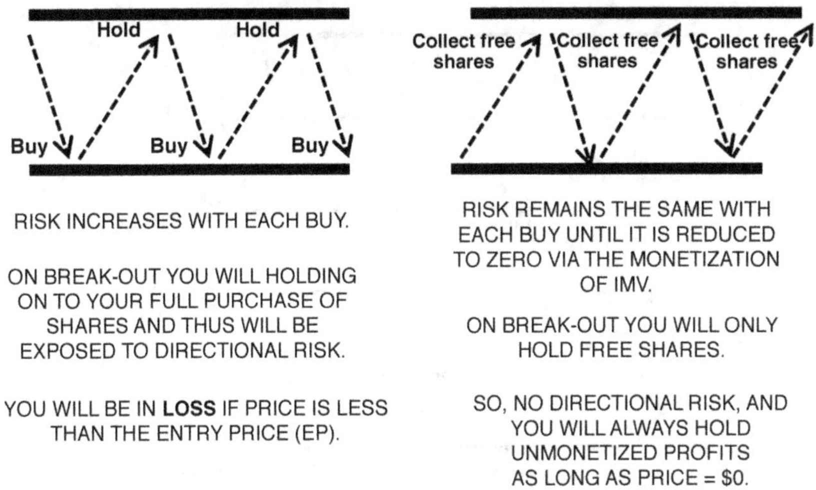

FIGURE 9.6 Conventional vs. Risk-Free Share Accumulation

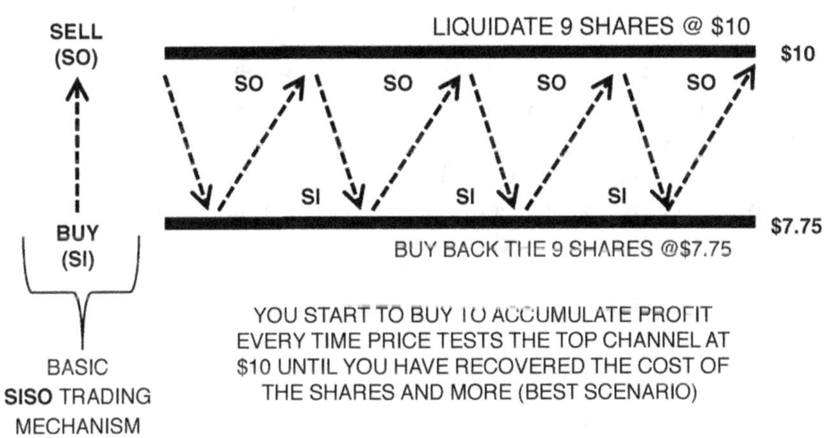

FIGURE 9.7 Accumulating Profit During Consolidation

So, every time price declines and hits \$7.75 you buy back the 9 lots. If you continue to do this, your CVAR will gradually decrease until finally, you may have 40 risk-free shares in the market (see Figure 9.7).

INTERVAL SLIP-THROUGH

Sometimes, vibrational mechanisms fail to secure vibrational returns in an oscillatory market as a consequence of interval slip-through, in which

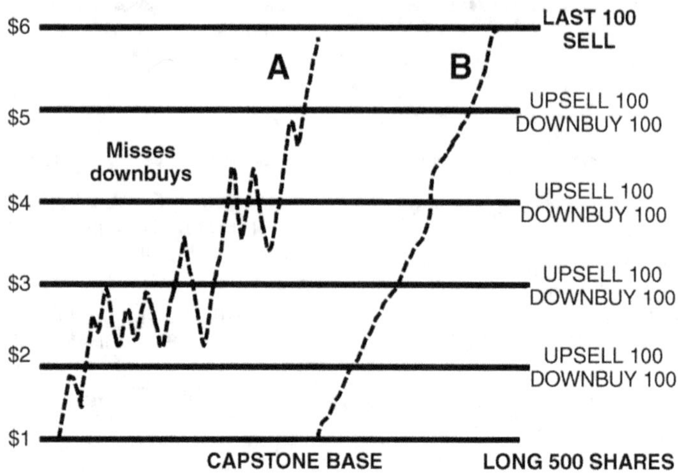

FIGURE 9.8 Slip-Through in Action

intervals pass without being filled. In order to generate maximum vibrational returns, downbuys must terminate at the upsells as many times as possible before price moves on to the upside or downside. Vibrational traders may reduce the interval size, thereby increasing the chances that markets oscillations will trigger multiple trades within the narrower price levels. But such an increase will also incur a higher total allocated working capital. See Figure 9.8 for the next example.

Compare vibrational profits derived from Price Path A and Price Path B.

Vibrational Profit from Path A = $1,500.

Vibrational Profit from Path B = $1,500.

We see that the returns are identical because of slip-through. One issue in vibratrading is that sometimes the price path only triggers the upsells and completely misses the downbuys, and the reverse can be true for downtrends. Sometimes even sideways markets fail to generate vibrational profits as the price intervals may be too wide to trigger the SISO or SOSI mechanisms.

Fortunately, this may be remedied via:

1. Interval Reduction I
2. Interval Reduction II
3. Market Selection.

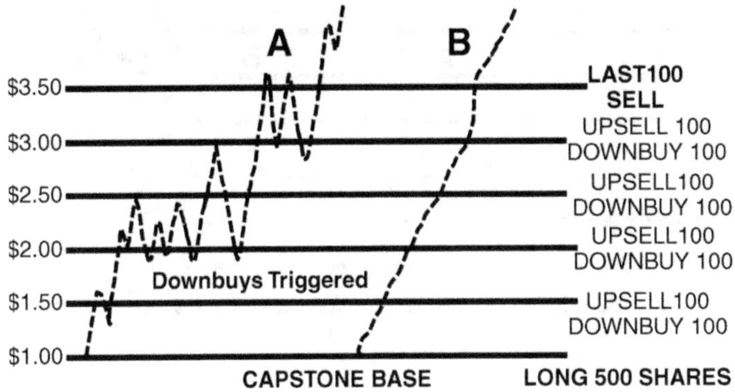

FIGURE 9.9 Less Slip-Through with Narrower Interval Sizing

Interval Size Reduction I

We can reduce the width or size of the intervals so that price oscillations are captured more frequently. The advantage of narrower intervals is the ability to achieve a more consistent income stream, resulting from more efficient vibrational extraction. The disadvantage is that the construct will require a higher total allocated capital in order to generate profits in a bounded manner. In Figure 9.9, the same Price Path A seen in Figure 9.8 can now generate more vibrational profits from the narrower price intervals.

Interval Size Reduction II

Interval Size Reduction II is essentially the same as Interval Size Reduction I. The sharesize, however, is reduced tremendously, providing the vibratrader with the opportunity to initiate more downbuy levels, thereby giving a much finer extraction of vibrational profits. There are significantly lower levels of return per oscillation compared to Interval Size Reduction I, which maintains the original sharesizes per level. One disadvantage of Interval Size Reduction II is that your full swing profits will decrease due to earlier exits of smaller sharesizes. The advantage is greater consistency of returns, since the extraction is much more frequent (see Figure 9.10).

Market Selection

We can also select stocks that are consistently more volatile to decrease slip-through, thereby increasing the ability of the vibrational construct to

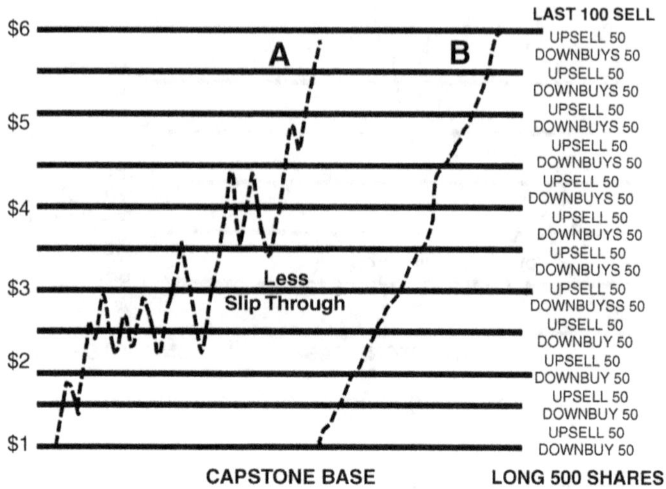

FIGURE 9.10 Interval Size Reduction II

capture price oscillations. We will discuss the issue of market selection in more detail in the chapter on volatility matching.

The disadvantage to this method is that more volatile stocks may sustain larger spreads. But the advantage is that historically volatile stocks should continue to remain volatile as part of their fundamental makeup. With more volatile stocks, gapping and erratic price moves may occasionally result in positive price slippage, especially with the use of entry and exit limit orders in vibratrading. Generally, fundamentals tend to drive volatility and technicals tend to drive trends (see Figure 9.11).

Narrowest Slip-Through Interval

It is imperative to know the narrowest interval size allowable. The limit or granularity factor is really about vibrational returns over trading cost. Your interval size should follow a simple condition:

$$\text{Interval Profit} = \text{Interval Size} \times \text{Sharesize} > \text{Trading Cost}$$

So, if your broker charges \$1.50 per 1,000 share transaction, that is, \$3 for one full round, then:

$$\begin{aligned}
\text{Interval Size} &= \$0.50\\
\text{Sharesize per level} &= 5\\
\text{Trading Cost} &= \$3\\
\text{Interval Profit} &= \$0.50 \times 5 = \$2.50 < \$3.
\end{aligned}$$

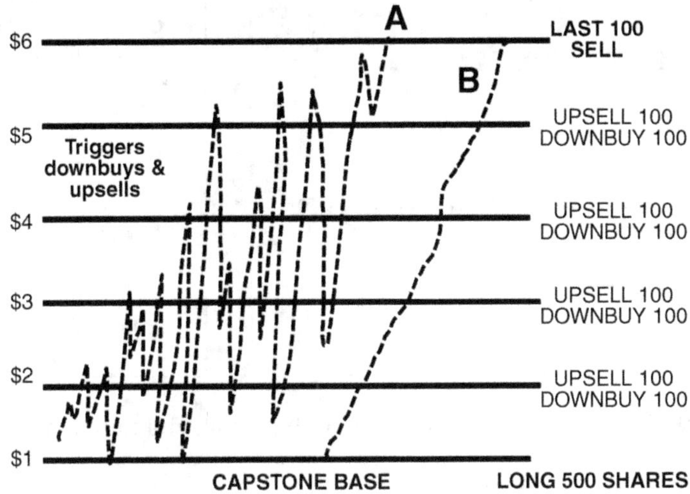

FIGURE 9.11 Selecting Volatile Markets Reduces Slip-Through

Since $2.50 is less than the trading cost, it fails to fulfill condition. Therefore, you will need to increase either interval size or sharesize. You may also try to reduce your trading cost by negotiating a better deal with your broker or finding a discount broker.

MACROSISO

The Macrosiso generally possesses a bottom-heavy structure since its minimum sharesize is at the capstone entry level and begins to scale in as price falls. All Macrosiso trades begin at the capstone entry level or from the apex, especially after a Macrososi full swing. Should price remain and oscillate within the pyramidal base, vibrational profits will continually be extracted from the market. This vibrational trading range extends from the apex or capstone base right down to the lowest possible price level at the floor of the pyramidal structure. The main difference between the Microsiso and the Macrosiso is that the Macrosiso withholds the upsell for two or more intervals beyond its downbuy.

One important metric in vibratrading is current value at risk, or CVAR, which is equal to the IMV less all monetized profits. Let us briefly refer to some diagrams to clarify and consolidate this concept, starting with Figure 9.12.

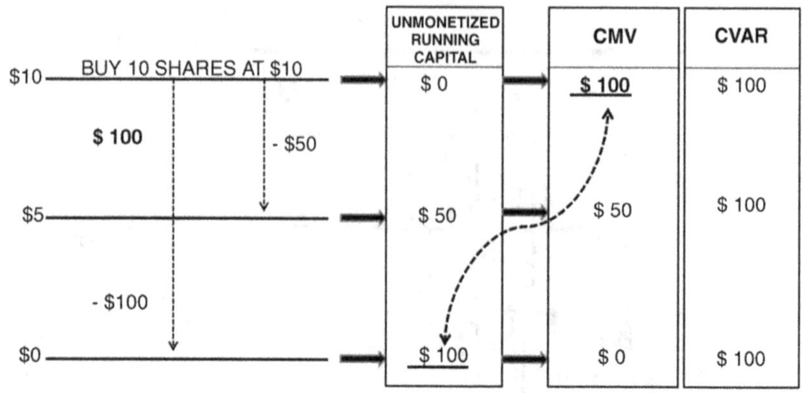

FIGURE 9.12 Relationship Between CMV, CVAR, and Running Capital

We see that:

Maximum Running Captial = CVAR = Value of Shares Purchased
= Initial Market Value (IMV)

Assume that we bought ten shares at $10, as per the diagram. When price declines to $5, our running capital increases to $50 and our CMV falls by $50. Regardless of where price is, the CVAR is constant. But what happens if we monetize some profits? To answer that question, we refer to Figure 9.13.

CVAR = IMV = Monetized $Profits

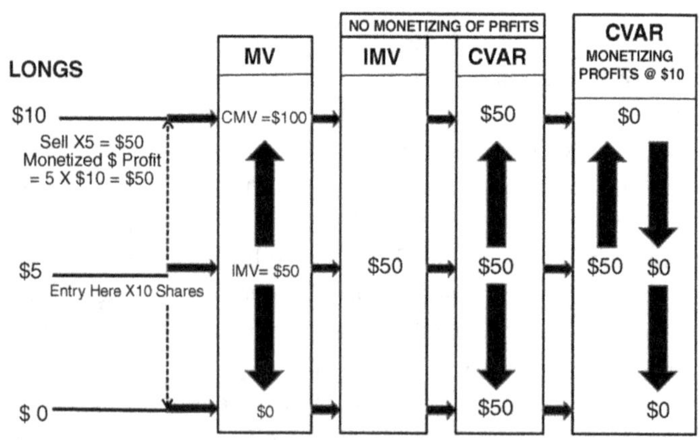

FIGURE 9.13 Definition of CVAR (Monetizing Profits)

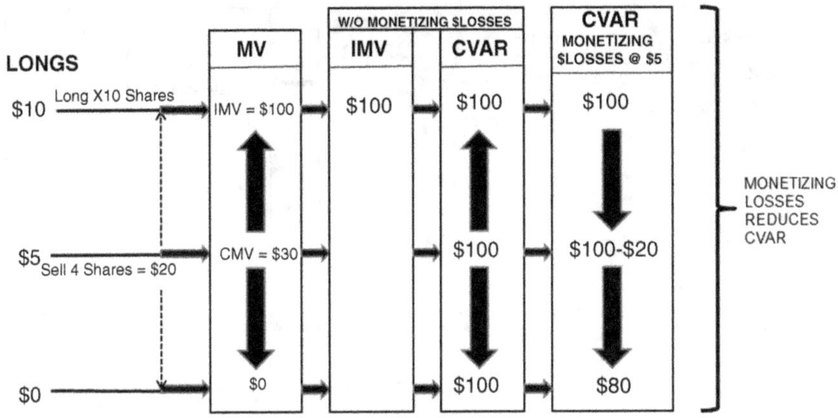

FIGURE 9.14 Definition of CVAR (Monetizing Losses)

In that example, we see that if we purchased ten shares at $5 and subsequently liquidated five shares at $10, the CVAR falls to zero, creating a risk-free trade. CVAR is therefore affected by monetizing profits and loss, as seen in Figure 9.14.

$$\text{New CVAR} = \text{Original CVAR} - \text{Monetized \$Losses}$$

Assume we bought ten shares at $10. We can clearly see from the example that liquidating four shares at $5 reduced the CVAR to $80 from its original value of $100. Therefore, CVAR is affected by the monetization of both profit and loss.

Before looking at the various vibrational Macrosiso constructs, we need to study the effects of three constructs to better understand the interplay between CVAR, IMV, and the break-even price, namely:

- Hedged Scale Construct
- Unhedged Scale Construct
- Cash Scale Construct.

Hedged Scaling Construct

A hedged construct is composed of long and short positions. The ratio of longs to shorts in a hedged construct must always be greater than one to maintain overall directional boundedness (see Figure 9.15).

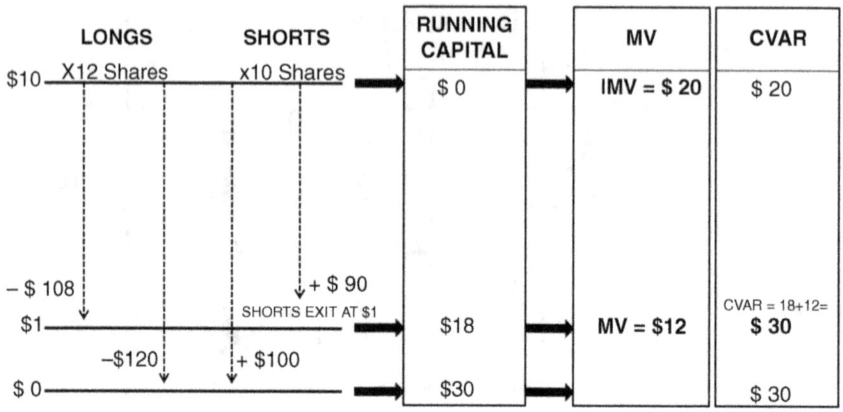

FIGURE 9.15 Hedged–Scale Construct

Calculating break-even point for monetized CVAR:

$$\text{Break-even } \$ = \$1 + (\$18/12 \text{ Shares}) = \$2.50 \text{ or}$$
$$\text{Break-even } \$ = \text{CVAR/Shares} = \$30/12 = \$2.50$$

Assume we long 12 shares at $10 and simultaneously short 10 shares at the same price level. We have a net long exposure in the markets. Price then declines to $1, whereupon we cover our 10 short positions. At $1 our CVAR is $30, after the $18 loss from the price decline and another potential $12 loss should price reach zero. Note that the running capital matches CVAR at zero. The break-even price is $2.50.

Unhedged Scaling Construct

Let us compare the hedged scale construct above to an unhedged scale construct, as seen in Figure 9.16. An unhedged construct is composed of only one type of positions. For the purposes of vibratrading, we shall examine only those with longs positions in accordance with the principles of directional boundedness.

We have essentially the same net long exposure. To maintain the same zero test level capital exposure, we now purchase 10 shares at $1. This maintains the risk between the two constructs, making them equivalent. From Figure 9.15, we notice that the CVAR is identical in value once the 10 additional shares are purchased at $1. The running capital also displays

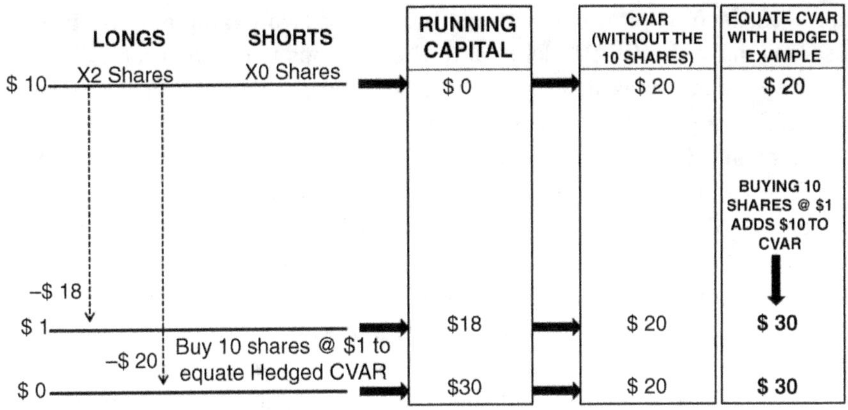

FIGURE 9.16 Unhedged-Scale Construct

an identical behavior to the previous model. Even the breakeven points are identical.

Let us compare this to the cash construct before we draw conclusions about the performance and characteristics of these three constructs.

Cash Scaling Construct

Refer to Figure 9.17 for an example of an "equivalent" cash construct.

In the cash construct, we delay buying until price declines to $1, whereupon we purchase 30 shares (so as to maintain the same level of

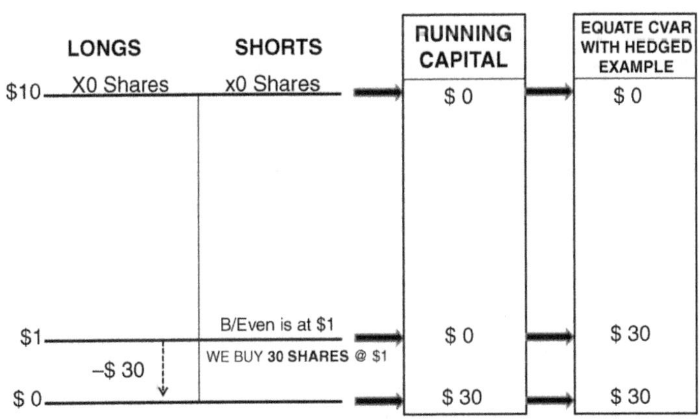

FIGURE 9.17 Cash-Scale Construct

zero test level capital exposure as in the last two examples) and in the process make the CVAR identical to the previous two constructs. But the running capital does display some differences, with the first drawdown only occurring below the $1 price level. There is a marked improvement in break-even price level.

The cash construct has the greatest earning potential since we can purchase 18 more shares for the same CVAR. It also has the lowest breakeven price level of the three. It experiences no running capital, margin, or trading costs until price tests the $1 level. The greatest drawback of the cash construct is opportunity risk. In the event that price fails to test $1, the cash construct trader may miss out on potential upside profits.

As far as the hedged and unhedged constructs are concerned, the unhedged construct has the advantage as it incurs much less margin during the initial period and the positions are much easier to handle. There is also no risk of a shorts recall. The hedged 12×10 has similar trend profit potential to the 2×0 unhedged construct; both are effectively long 2 lots. It only makes sense to implement a hedged construct once you have already decided on the lowest trend exposure lotsize, as a method to gain extra short profits in the process.

These examples show that cash trading is a form of passive trading. Doing nothing and waiting in cash is the most efficient and cost effective way to trade, despite the possibility of missed trading opportunities. You sometimes really do get more bang for your buck—as they say, cash is king!

EXTRACTING MACROSISO VIBRATIONAL PROFITS

Since we already understand that a Macrosiso mechanism upsells a down-buy across two or more intervals, we must now decide at which upsell to exit, the number of upsells exiting simultaneously, and how to ascertain the amount of profit prior to the exit. To do so, we need to alter scale factor, share size, and share multiples.

Scale factor is the ratio of the number of shares between an upsell and its immediate lower downbuy. Share size is the actual amount of shares at any scaling price level. Capstone shares are the actual number of shares at the capstone entry level. Share multiples is the ratio of share size over capstone shares.

These come into play when determining modes of exit from a Macrosiso downbuy.

Vibrational Constructs and SISO Exits

There are two basic exits from a trade within the vibrational structure:

1. Cost-Based: The removal of running capital at exit.
2. Profit-Based: The yielding of a profit at exit.

In addition, the profit made or cost removed at the exits will be based on either of the following:

1. Upsell Termination: Reduction or removal of running capital, or the yielding of profit based on the upsell level.
2. Downbuy Termination: Reduction or removal of running capital, or the yielding of profit based on the downbuy level.

Modes of exit from a Macrosiso downbuy play off these themes. One option is the removal of all or partial running capital in one upsell. You may also remove running capital through several upsells or yield a profit over one or a specific number of upsells.

If running capital is expected to be removed in one upsell, then the upsell must be at the average price level. For profit, the upsell must be higher than the average price. One great advantage of removing all running capital is that it gives you the opportunity to exit the ETF or CFD at the upsell, since the CMV of all positions is equal to or higher than the IMV.

Some of the parameters alterable include interval size, the number of scale intervals across the effective range, scale factor, and capstone share size. Some of the selectable parameters are PLR, the effective trading zone, and the instrument or stock traded.

Cost-Based Constructs

The cost-based construct is designed to remove all running capital in a single upsell, regardless of where the upsell occurs. Every upsell will automatically remove all running capital associated with the construct at that time, as shown in Figure 9.18.

If we use L0 to represent the capstone base, and L1 for the level below and so on, here we see that the scale factor (SF) is two from L2 downward.

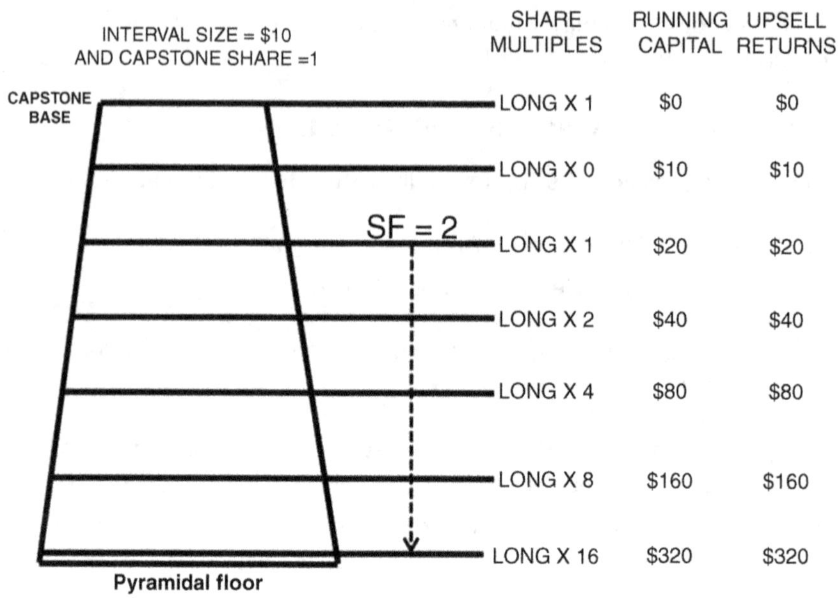

FIGURE 9.18 Cost-Based Construct

The running capital also increases with a scale factor of 2. As price declines, we see the running capital increase along with the returns, which match it perfectly. In effect, the construct removes all running capital no matter which upsell is triggered.

In terms of process, we see that the share multiples rise at each successively lower downbuy level. Since the capstone sharesize is 1, this means that the actual sharesizes are equivalent to the share multiples. If the capstone sharesize were 10 shares, then the actual sharesizes in this example would be 10, 0, 10, 20, 40, 80, and 160; the actual downbuy sharesizes are the product of the capstone sharesize and share multiples.

At L2, or the third level from the capstone entry level, we see that the next higher running capital is $10. Since the sharesize at L2 is 1 share, this would mean a return of 1 share × $10 interval size, or $10 on the upsell. This return will remove the $10 running capital associated with that level, thereby lowering the cost basis to zero for holding shares there. Note that we still have working capital in the construct. If no shares are sold at each upsell, the CMV will equal the IMV of all shares within the construct.

In another variation of the cost-based construct, the L1 sharesize is increased to one to allow for vibrational profits at every level. The upsells

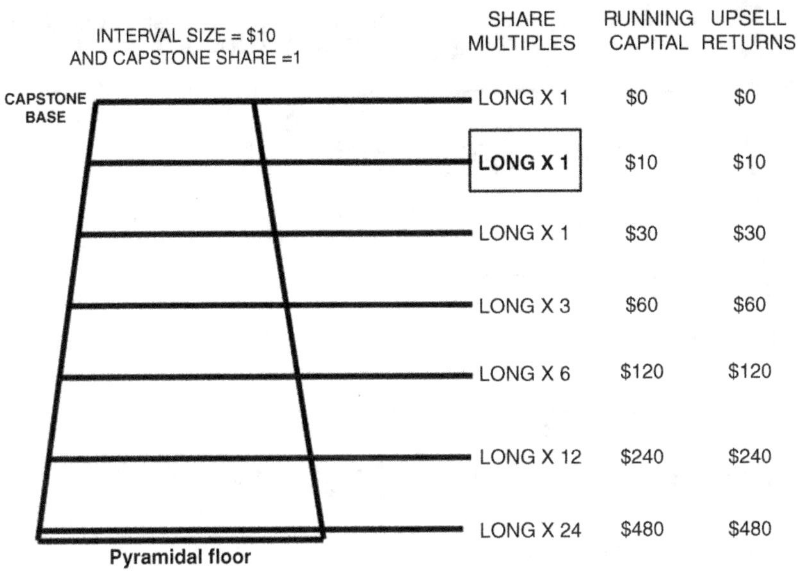

FIGURE 9.19 Cost-Based Construct (Fully Vibrational)

perform almost identically except for the slight change in sharesizing (see Figure 9.19).

Profit-Based Constructs

The first profit-based construct is designed to yield one unit of profit on termination at the upsell level.

As we can see from Figure 9.20, not only did each upsell remove all running capital, it also yielded a vibrational profit of $10. If price continues to move to the upside, additional upsells will produce the same $10 return right until price reaches the capstone entry level. In fact, vibrational traders try to hold off the upsells until price has risen considerably in the hope of extracting a greater return over a larger number of intervals. In the example above, we could easily arrange the downbuy sharesize so that the construct would yield any number of units of profit per upsell.

Phi-Based Constructs (Martingale)

The objective of this methodology is to orchestrate full recovery of all running capital—just like in the profit based construct—during a Macrosiso

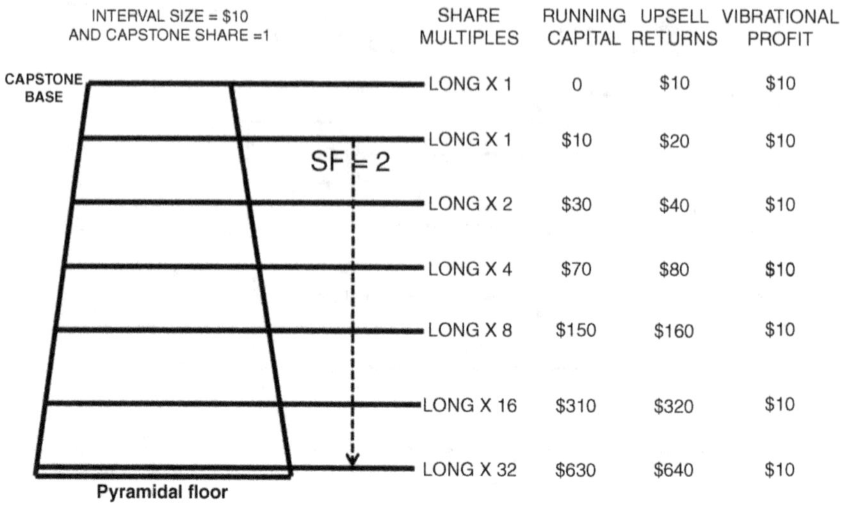

FIGURE 9.20 Profit-Based Construct

trade with just one upsell. It trades within the pyramidal base, and in the capstone after a full swing. This vibrational trading technique works best with fewer scale intervals as the capital requirement could escalate rapidly. Ensure capital sufficiency before embarking on phi-based trading. You can always reduce sharesizes to ensure additional affordability.

With the phi-based or Martingale-based construct, running capital based on the downbuy level is removed after one upsell. The method is similar to the cost-based and profit-based constructs, where the oscillation terminates at the upsell instead. The scale factor required to achieve this is strangely part of the Fibonacci sequence.

Dividing the share multiples within the series gives a scale factor equal to the universal number phi (φ), or 2.618. This ratio is closely connected to the Fibonacci series and has been used throughout history. Can you see the Fibonacci series evident in the share multiples in Figure 9.21? The numbers 1, 3, 8, 21, 55, 89, 144 and 377 are part of the Fibonacci sequence. If you divide 377 by 144 and so on as the series expands, you will get asymptotically closer to 2.618.

I will use this illustration to explain what the downbuy termination is all about. Functionally, it is not unlike the upsell termination in the profit- and cost-based constructs. The upsell is still the main determinant of the

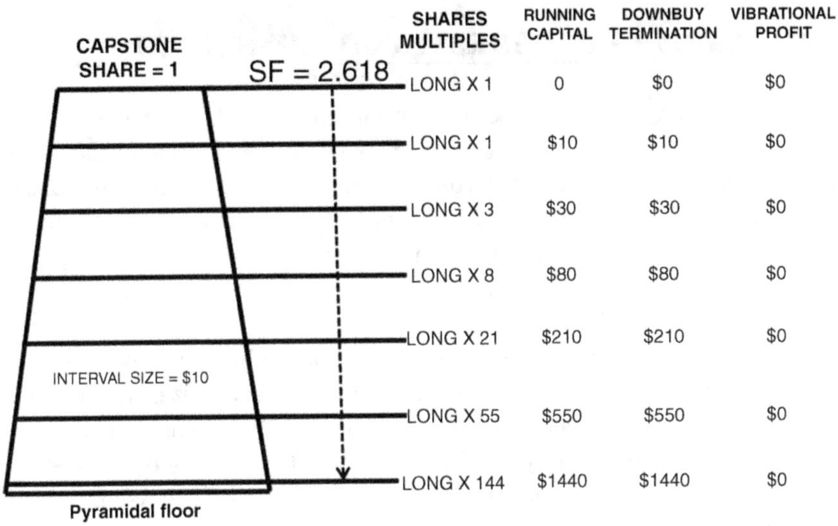

FIGURE 9.21 Phi-Based Construct

returns generated, but the amount of shares required to upsell is based on the running capital at the downbuy price level.

Let's focus on the sharesize at L3, which is 8. We observe that the running capital at L3 is $80, therefore, we upsell 8 shares. This would usually generate $50 profit as the running capital at L2 is only $30. But in downbuy terminations, the trader assumes that price will return to the downbuy level and stay there indefinitely. The idea is to remove all running capital at the downbuy level in the event that price does revert back to this lower level.

You may use the upsell within the phi-based construct as an exit strategy for ETFs or stocks. As you know, you cannot exit unless you can do so without depleting capital. The phi-based construct allows you to exit your trade unit at the upsell because it is higher than your break-even exit. Exiting below the average or break-even price is an unbounded activity and will result in capital depletion.

We shall be looking at the interaction of various scaling parameters for the phi, profit, and cost constructs in a later section. This section is to introduce the concept of the various Macrosiso formulations. Remember, you can create any manner of construct to extract vibrational returns from the markets. You need not base your formulations on just one of the three models presented.

THE "ARBITRARY" VIBRATIONAL CONSTRUCT

The way a vibrational construct is built largely depends on the amount of disposable capital available, the desired speed and magnitude of the vibrational returns expected, as well as other factors like the effective range of the traded instrument. All these factors determine the efficiency and performance of the vibrational construct, but the most influential factor is the amount of capital available. Generally, the more capital invested, the better the performance of the construct.

An arbitrary construct basically means that the builder may choose to alter any aspect of a construct. The scale factor, interval size, and effective range of the construct need not be fixed and may in fact change according to the degree of price volatility, market range, and profitability of the system.

As far as scale factoring is concerned, the construct may be:

- SF = 1 (Equal sharesize allocation throughout construct)
- Top Heavy (Higher sharesize allocation at top end of construct)
- Bottom Heavy (Higher sharesize allocation at bottom end of construct)
- Center Light (Higher sharesize allocation at either end of the construct)
- Center Heavy (Higher Sharesize allocation at center of construct)
- Alternating Light or Heavy
- Center Void (No sharesize allocation at center of construct)
- Volatility Dependent (ATR-Driven scale factoring, etc.).

As far as interval sizing is concerned, the construct may have:

- Narrower and more concentrated intervals at the top of the construct
- Narrower intervals at the bottom of the construct
- Narrower intervals at the centre of the construct
- Narrower intervals around the current price action (see Range Zoning)
- Interval size equally spaced throughout the construct
- Interval size varying with volatility.

Unbounded Macrosiso Vibrahedging

Unbounded macrosiso vibrahedging helps traders maintain a vibrational construct with reduced capital outlay, while boosting potential profits

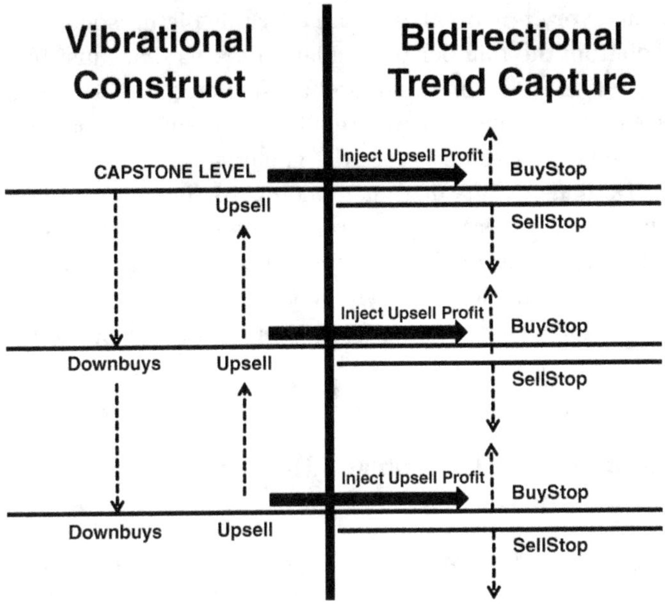

FIGURE 9.22 Unbounded Macrosiso Vibrahedging

should the markets move favorably. The strategy is basically two methodologies in one. On one side you have the vibrational constructs extracting oscillational returns, and on the other you have a bidirectional trend capture mechanism (see Figure 9.22).

The entire process only begins once some vibrational profits have been generated by the upsell action within the construct. Let us assume that we are trading the silver market with 100 troy ounces and an interval size of 400 cents. The upsell at the capstone entry level resulted in a return of $400. Since this is "income," we may use it immediately. We deploy this $400 return in the bidirectional mechanism by constructing a straddle with both buy and sell stops, hoping to catch either an upside or downside breakout. We allocate each stop order with 100 ounces and a straddle width of 20 cents. In other words, the stop loss for each stop entry order is 20 cents.

The idea here is that if the break-out is successful, the trader will benefit from the extra trend profit made from the bidirectional mechanism, which will offset or hedge the running capital as price declines. But if the break-out is to the upside, the trader makes double the profits: from both the 100 ounce long position within the vibrational construct and the 100 ounce long position held in the bidirectional mechanism.

In reality, very few break-outs are ever this picturesque. But there is one consolation: the number of oscillation losses that must be endured before the entire $400 vibrational profit is depleted is 400 cents/20 cents, or 20 losses. This means that we are allowed to suffer about 20 failed or non-performing break-outs before the game is up.

One advantage of this dual methodology is that it gives FOREX and futures directional traders a chance to implement vibrational trading within what is typically a trend trading environment. Finally, if the working profit of $400 is somehow depleted, the vibrahedging mechanism will not be able to restart its hedging function until a new upsell profit is generated by the vibrational construct.

The Unbounded "Dirty Harry" Method for Position Allocation

Dirty Harry is an unbounded technique for reducing the amount of capital outlay required to maintain a vibrational construct, especially if price accesses the lower levels of the construct. As lower downbuys are triggered, more running and working capital is required to fund purchases at these lower levels. The unique name for this technique was suggested by one of my graduates during a Vibrational Trading Masterclass webinar back in 2008.

One reason to use this unbounded technique is a larger-than-expected price decline. Knowing that there is insufficient capital at hand to fund further downbuys or maintain all current long positions to zero test level, the trader may decide instead to jump ship. A quick and "dirty" way of reducing the amount of capital required would be to exit a number of positions at a predetermined price level via the capital depleting stop loss. Should price later rise back beyond this level, the vibrational trader may re-initiate the positions that were exited. It is an unbounded methodology, since there will be capital depletion caused by oscillation losses at the point of exit and re-entry. As a result, this technique should be used only as a last resort (see Figure 9.23).

Recouping Running Capital in a Vibrational Construct

There may be times when you are required to recoup all running capital over an arbitrary interval size (see Figure 9.24.)

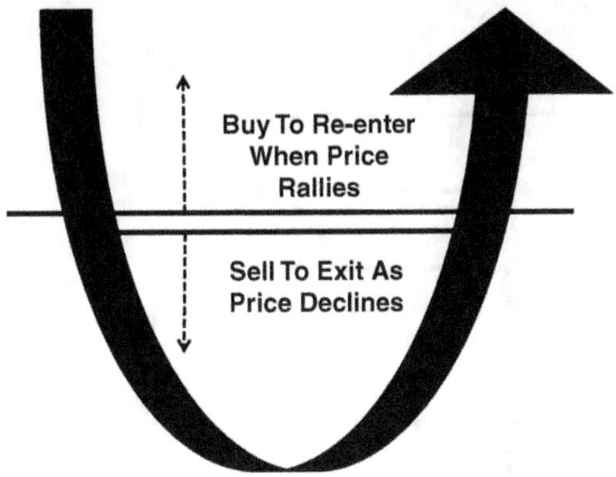

FIGURE 9.23 Capital Depletion at Re-entry

Assume that the total running capital is \$100 at the predetermined upsell level and the arbitrary interval size is \$3.33.

$$\text{Sharesize} = \text{Total Running Capital at the Upsell Level/Interval Size, or}$$

$$(\$100/\$3.33) = 33 \text{ Shares.}$$

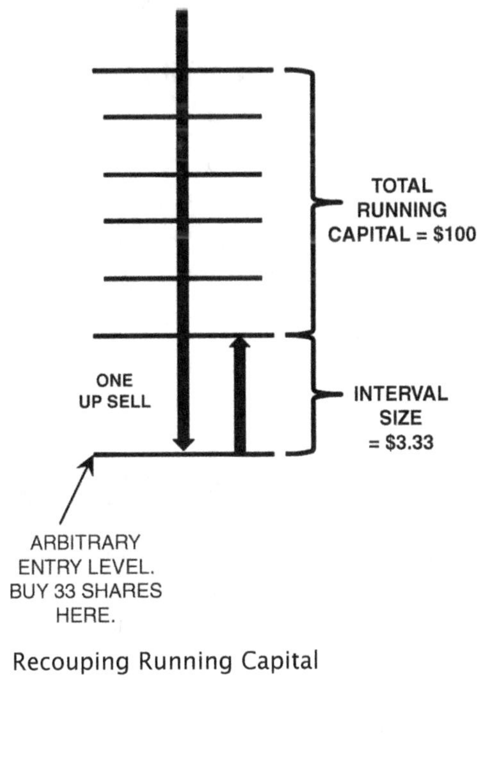

FIGURE 9.24 Recouping Running Capital

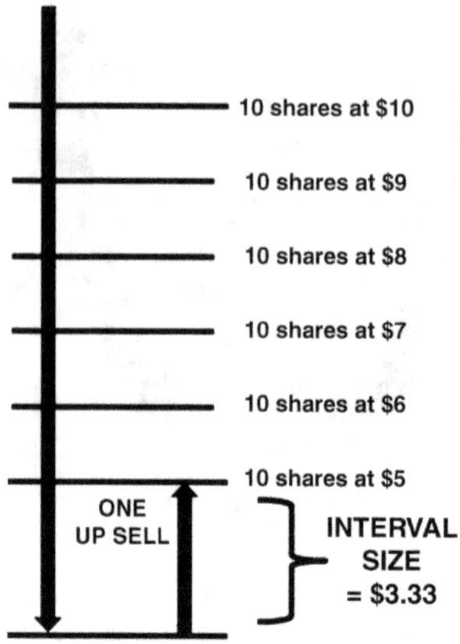

FIGURE 9.25 Recouping Working Capital

Therefore, if we buy 33 shares at the arbitrary level, we would be able to recover the total running capital in a single upsell.

Recouping Working Capital in a Vibrational Construct

Let us assume the same set-up as the previous example. But now, we want to recoup the total working capital within a single upsell (see Figure 9.25).

$$\text{Sharesize} = \text{Total Working Capital at the Upsell Level/Interval Size.}$$
$$(\$100 + \$90 + \$80 + \$70 + \$60 + \$50)/\$3.33 = 135 \text{ Shares.}$$

Therefore, if we buy 135 shares at the arbitrary level, we would be able to recover the total working capital in a single upsell.

UPSIDE BOUNDED MACROSISO AND MICROSISO

The upside bounded macrosiso is unique in that it is both a vibrational and a trend capture mechanism. The upside Macrosiso can initiate an entry anywhere from the pyramidal base to just below the apex. It can do this because the upside Macrosiso has already taken into account the total allocated capital required to fund all downbuys to just below the apex. Price needs to first cross above the downbuy level. Then, upon a quick retest, the downbuy is entered. We do not use buy stop entry orders to initiate the buy levels in an uptrend, but we still maintain the use of buy limit orders to trigger the downbuys to preserve order entry boundedness. Once the longs are initiated, the positions may be exited across a single interval (Upside Microsiso) or multiple intervals (Upside Macrosiso). Only limit orders are used.

UNBOUNDED UPSIDE MACROSISO MECHANISM

In the unbounded upside macrosiso mechanism, longs are entered via a buy stop order as price rises. Therefore, there is a possibility that long entries may experience negative slippage or gapping action, becoming unbounded. Nevertheless, the mechanism must maintain some degree of capital boundedness by ensuring that the funding of all higher long entries is accounted for within the construct. Once the longs are initiated, the positions may be exited via a sell stop order across two or more higher intervals.

UNBOUNDED HEDGED VIBRATIONAL CONSTRUCTS

The unbounded hedged short rider and scaler is not only about reducing downside capital costs, but also extracting both long and short vibrational profits. It is an imperfect hedge. The shorts have a smaller position size than longs at all times as the rule of directional boundedness requires a net long position.

It is critical to note that this methodology is only utilized on the initial condition that the vibrational trader has decided on the maximum capstone IMV, thereby implying a fixed amount of profit drivers at the capstone entry level. In the following examples, we shall be using the phi-based construct as our vibrational model. Of course, the vibrational trader is free to fabricate any manner of construct to fulfill this objective.

In the short rider the shorts exit arbitrarily, only to re-enter at higher prices, if possible. The vibratrader scales out to the downside at each level via predetermined exits.

As seen in Figure 9.26, additional short vibrational profits may be generated along with long vibrational profits via the short scaler. It is best

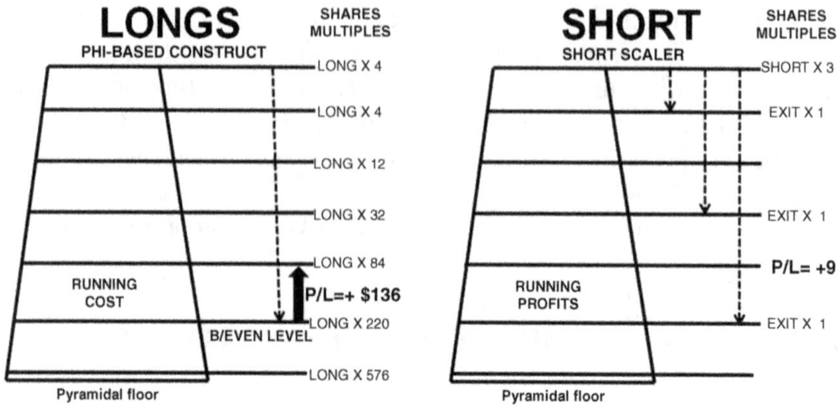

FIGURE 9.26 The Short Scaler

to re-establish new short scaling positions one or two intervals above each short exit. This will create a vibrational zone where long and short returns are generated. As you will learn later, the short scaler is a simplified version of the big hedge.

Similarly to the short scaler, the short rider allows additional short vibrational profits to be generated (see Figure 9.27.) It is very important to re-enter a short rider position higher than the last short cover level. Of course, this will preclude any short entry if it stays below the previous short cover level. Using the capstone entry level as a possible re-entry is recommended. For more advanced bidirectional profit extraction, read the chapter on vibrabidirectional methods.

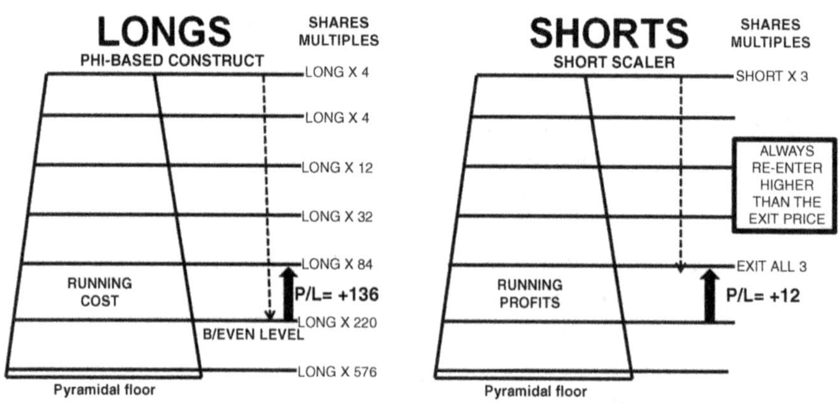

FIGURE 9.27 The Short Rider

CHAPTER 10

Diversification in Vibratrading

We have learned that all entries into the pyramidal construct must be capstone-based to be bounded, except for the Unbounded Upside Macrosiso. You will soon learn to combine micro and macro trading in order to form a more efficient vibrational construct.

But, no matter how well we design the constructs to extract trend and vibrational profit, there is no way to prevent via the deployment of vibrational mechanisms alone the collapse of the pyramidal foundation. See Figure 10.1 for an explanation.

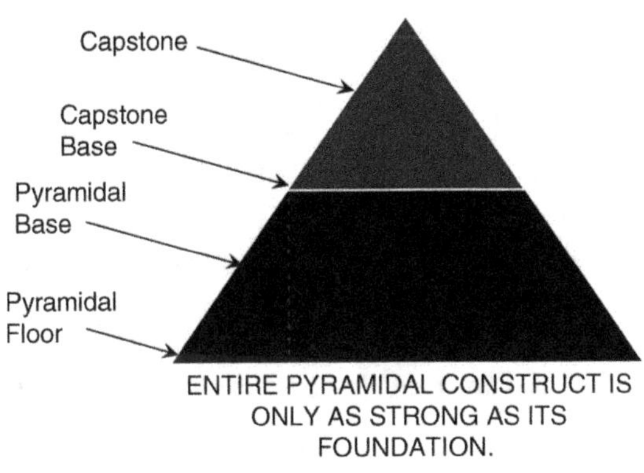

FIGURE 10.1 The Pyramidal Structure

The only way to neutralize or at least drastically reduce the possibility of a pyramidal foundational collapse is through diversification. We have six levels of diversification to ensure that our pyramidal foundation is as zero test level–proof as possible.

BOUNDED VERSUS UNBOUNDED ZERO TEST LEVEL EVENT

There are three main factors which help strengthen the foundations of a pyramidal construct:

1. Demand creates a powerful upward force, especially in undervalued stocks. Demand works so well (and even better near zero test level) because of the extremely attractive PLR, BLR, and MLR. This natural support for price creates phenomenal retracement or rebound profits.
2. Capital risk (CVAR) is much less at lower entry levels. This allows investors to enter and buy stocks that they could not have otherwise afforded.
3. Objective Support Level means that all market participants have the same reference point for an absolute support, as further downside price excursion is not possible below zero test level. Therefore all participants will react accordingly to this very objective price extreme.

Refer to Figure 10.2. The capstone represents the entry phase of a trade, poised initially for trend profits. While price hovers above the capstone base, vibrational profits are continually extracted. Below the capstone base, further micro-vibrational profits are captured in addition to larger macro-vibrational profits. These vibrational profits grow greater as price approaches the pyramidal base, because of increasing PLR, BLR, and MLR. See Figure 10.3 for a graphical representation.

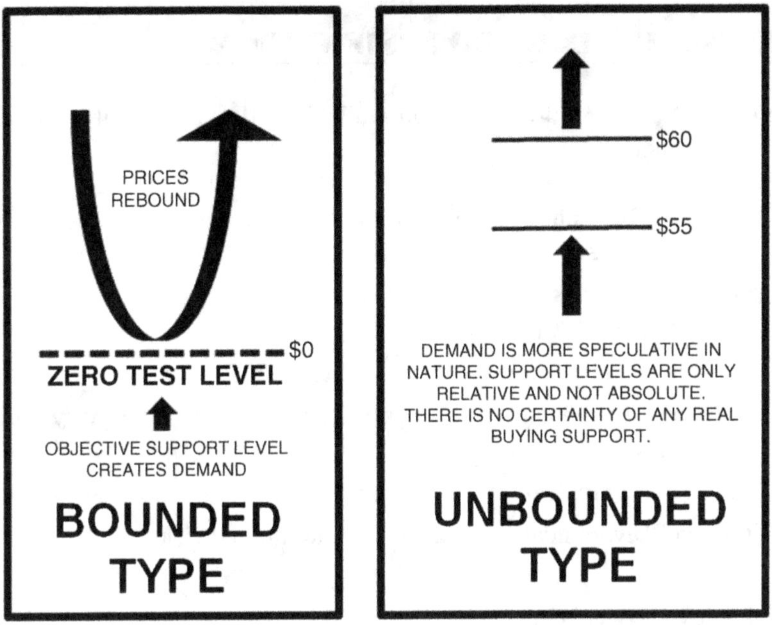

FIGURE 10.2 Absolute vs. Relative Support

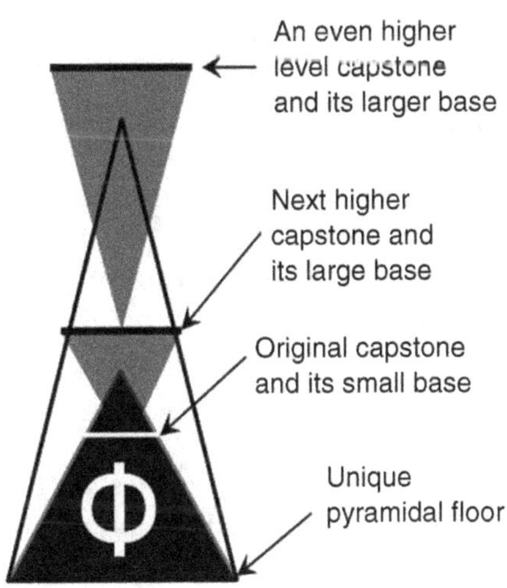

FIGURE 10.3 Unfolding Pyramidal Structure

THE SIX LEVELS OF DIVERSIFICATION

There are six levels of diversification available to the vibrational trader and investor, namely:

- 1st Level Diversification: ETF's stock replacement mechanism and price averaging
- 2nd Level Diversification: Larger basket of component stocks/ commodity ETFs/CFDs
- 3rd Level Diversification: Owning multiple ETFs
- 4th Level Diversification: Owning multiple ETFs in negatively correlated markets
- 5th Level Diversification: Oscillatory propagation technique
- 6th Level Diversification: Currency denominated risk.

1st Level Diversification

This is achieved simply by owning a single equity ETF. An ETF reduces specific risk and softens the adverse effects of systemic and systematic risk. It also strengthens the pyramidal foundations via its stock replacement mechanism and price averaging. It is still superior to owning single stock, which has no mechanisms preventing share prices from falling to zero.

2nd Level Diversification

Owning a larger component equity ETF provides an even stronger protection against a zero test level event, especially with the stronger demand as it approaches zero. This strengthens the pyramidal foundations more than ETFs with lesser component stocks.

3rd Level Diversification

Diversifying into more than one ETF provides even stronger protection against the total allocated capital being lost during a zero test level event. This forms an incredibly strong foundation for further equity risk management of the account.

4th Level Diversification

The vibratrader ensures that all capstone-based trades are balanced with pyramidal-based trades across multiple ETFs. Through the careful use of a negatively correlated market, Macrosiso and Microsiso trades are compensated—and somewhat funded—by the profits captured via the Macrososi and Macrosimo trades. This further reduces systematic or market risk.

We need to locate negatively correlated markets in order to implement an effective Level 4 diversification. Unfortunately, correlation between markets occasionally breaks down due to decoupling as a result of fear and flights to safety, or decoupling as a result of timing delays in response to market action.

Some correlations are more reliable as they are based on more of a structural correlation, as opposed to correlations driven by natural market forces. Below are some examples of negatively correlated markets driven by natural market forces. It must be noted that these negative correlations work much better in the absence of all decoupling effects:

- Oil vs. USD
- Commodities vs. USD
- Equity Markets (Indices) vs. USD
- Bonds vs. Commodities.

Structurally driven negative correlations are much more reliable and consistent. One clear example is the UUP (PowerShares DB US Dollar Index Bullish Fund), which is constructed with the Euro as a major weighted component in its calculations. Hence, trading the FXE (Currency Shares Euro Trust) against the UUP will, to a large degree, help ensure a more consistent negative correlation (see Figures 10.4 and 10.5).

As we can see from the charts, structurally driven correlations are very consistent in maintaining the desired intermarket relationship. Figures 10.4 and 10.5 display consistent inverse market action. Let's compare this to market-driven correlations.

There is a natural market-driven negative correlation between the GLD ETF (SPDR Gold Trust Shares) and the UUP Fund, but as you can

FIGURE 10.4 FXE (March–December 2010)

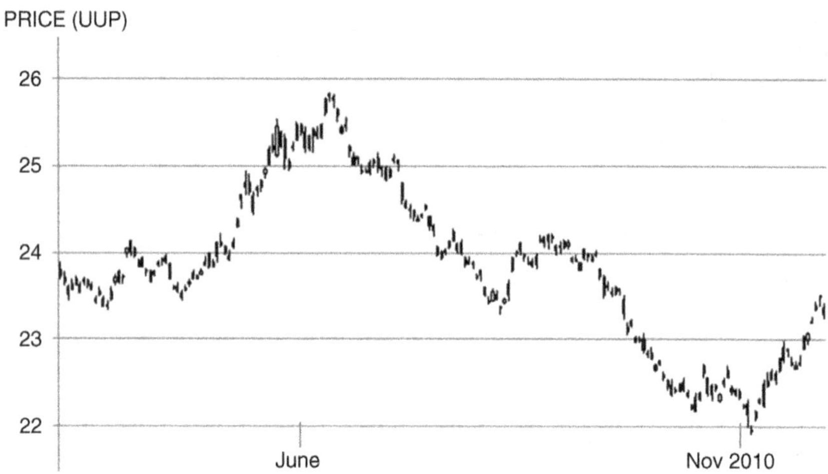

FIGURE 10.5 UUP (March–December 2010)

see from Figure 10.6, it is not as reliable as the structurally-driven correlations. Its inverse relationship to the UUP is less consistent than that of the FXE. This is because there is no mathematical basis driving interaction between these ETFs.

It is important to maintain optimum balance during inverted swings. You will need to decide on the capstone IMV for each ETF, including all capital conversion possibilities. Adjust the scaling so that the upside profits

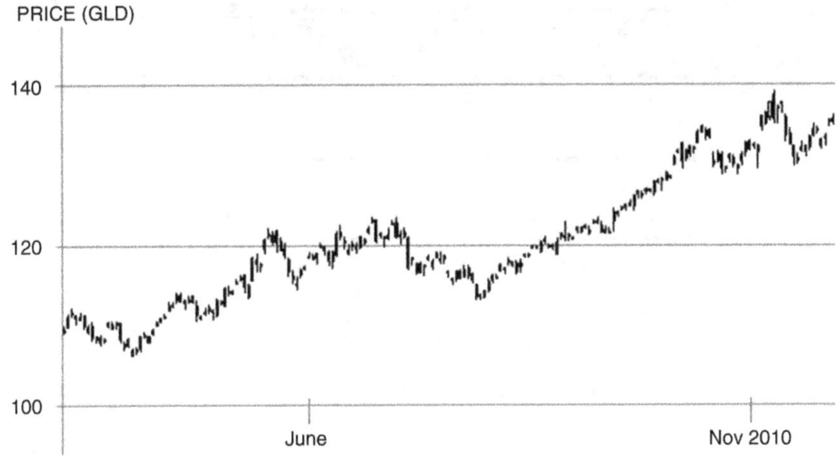

FIGURE 10.6 GLD (March–December 2010)

minimize the running capital between the traded ETFs during the negatively correlated, or inverted, swings. A full cycle inverted swing will yield vibrational profits from both negatively correlated ETFs.

There may be marked differences in PLR and BLR between many of the ETFs involved in the negatively correlated trades. This creates a kind of PLR spread and requires some rebalancing. Experiment with the scale factor, sharesizes, and interval sizing to help balance out the PLR spread problem, so that the potential earning power of ETFs with a low PLR matches that of the higher PLR ETFs. It will not be a perfect or easily implemented solution, but it does go a long way toward maintaining a more balanced intermarket exposure. You may also use the short rider and scaler strategies.

5th Level Diversification

The oscillatory propagation technique helps to generate vibrational returns at levels around the capstone entry level, though it need not be specifically used in this manner. This technique actually arose from requests for a way to trade stocks vibrationally, rather than trading and investing in ETFs. Though stocks have no mechanism to prevent share prices from falling to zero, the oscillatory propagation technique is as close as you can get to extracting vibrational profits in an unbounded trading environment.

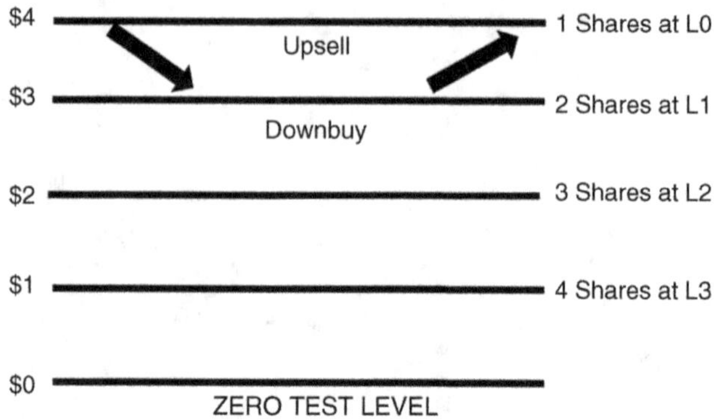

FIGURE 10.7 Oscillatory Propagation Technique

Oscillatory propagation works as follows. Let's assume that you are interested in extracting vibrational returns from some stocks. Let us also assume, for the sake of convenience, that you bought one share at \$4, two shares at \$3, three shares at \$2, and finally four shares at \$1 (refer to Figure 10.7).

First you calculate the total allocated capital for a single stock, (or average total allocated capital in the real world!) You want to find out how many stocks it would take to establish a self-financing situation whereby you would be able to purchase a new stock for free after a single vibration return. The vibrational upsell return from L1 will be \$2 in this example.

$$\begin{aligned}\text{Total stocks required} &= \text{Total allocated capital of sample stock/}\\ &\quad\ \text{single interval return}\\ &= (\$4 + \$6 + \$6 + \$4)/\$2\\ &= 10\,\text{stocks.}\end{aligned}$$

This means that for the type of stock that you are interested in, you will need to purchase approximately 10 stocks to be able to purchase a similar stock for free every time price oscillates at the top end of the trading range. Of course, if price oscillates at a deeper price level like L3, you would be able to purchase two stocks for free (\$40/\$20) with every oscillation.

We can now see that as long as the oscillation creates stocks at a faster rate than the stocks wind up, the system will self-finance new purchases. It is this feature that helps protect against specific risk when dealing with single stocks.

6th Level Diversification

Finally, we must guard against any capital erosion via currency risk. We can somewhat reduce the effects of currency risk by diversifying our portfolio into other stable currencies. Unfortunately, this does not protect our portfolio against inflation risk. We can, to a degree, hedge against inflation by investing in commodity-based ETFs.

CHAPTER 11

Volatility Matching

Matching stock volatility to the appropriate vibrational construct will yield greater profitability in the long run. This is an important technique because it helps to generate a greater profit with less total allocated capital. Let us now examine the three volatility matching techniques.

There are three main volatility setups:

1. Historical Range Volatility (natural volatility)
2. Event Trading (high volatility)
3. Range Zoning (medium to low volatility).

HISTORICAL RANGE VOLATILITY (HRV)

Look for a high daily volume ETF. It should display an HRV with high PLR and BLR and be actively volatile throughout most of the range. You can use the ATR indicator to gauge the average daily or weekly volatility, then customize the scaling intervals according to this value. Commodity ETFs and some ETFs with fewer component stocks are good candidates for volatile behavior. (Also note that many leveraged ETFs have very large daily ranges, but be mindful of the daily reset feature, unless you are rebalancing the sharesizes daily.)

To locate volatility in the market, look for high daily, weekly, and monthly ATR and stock beta value. You should also consider the range activity for each of these periods. You may use any popular scanning software or manually perform the scan by eyeballing charts of typically high volume stocks.

EVENT TRADING (HIGH VOLATILITY TRADING)

Look for major economic events and announcements, like reports, meetings, or earnings.

For best results, the ETF should have a high PLR and BLR. Suitable events include major economic reports related to housing (starts, home sales); GDP; inflation data (CPI, PPI); unemployment (NFP, initial claims); and especially inventories. Though ETFs do not have earnings reports, the major reports from some of the larger component stocks may cause some volatility in the ETF itself. You initiate a capstone base entry prior to the report and trade the post-event volatility vibrationally.

RANGE ZONING (MEDIUM TO LOW VOLATILITY TRADING)

Look for narrow consolidating price patterns. For best results, the ETF should have a high PLR and BLR. A good candidate would have been the Financial Select Sector SPDR ETF, which was in a narrow consolidation from April 2009, as evidenced in Figure 11.1.

This was ideal for range zoning as there could also be profits should commodities explode to the upside with Macrosimo entries.

Range Zoning Method 1: The Macrososi Trade

Let us assume that your maximum capstone IMV equals $2000.

This value has taken into consideration that you may need to execute Micro- or Macrosiso trades should price fall. You intend to begin a

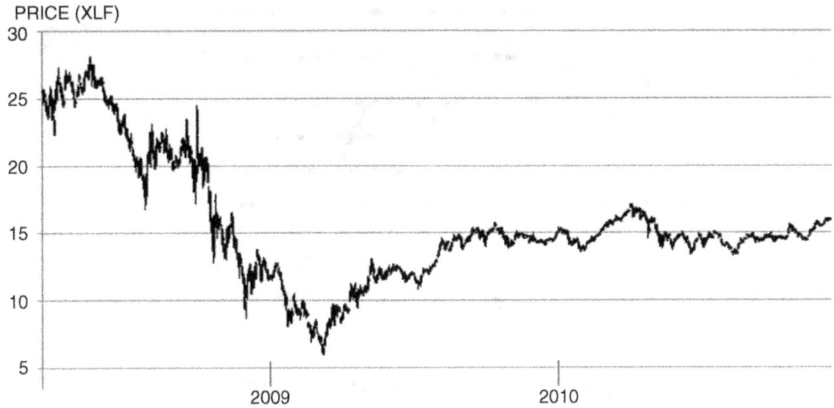

FIGURE 11.1 XLF ETF (March 2008 - December 2010)

Macrososi trade by zoning in at the \$10 capstone entry level, with five levels at an interval size of \$2 each within the capstone base.

$$\begin{aligned} \text{Scale Factor} &= 1 \\ \text{Entry Price} &= \$10 \\ \text{Interval Size} &= \$2 \end{aligned}$$

Therefore the Macrososi upper zone limit, or apex, $= \$10 + (\$2 \times 5) = \$20$.

$$\begin{aligned} \text{Shares Available} &= \text{Capstone IMV/Entry Price} \\ &= \$2,000/\$10 \\ &= 200 \text{ shares.} \end{aligned}$$

$$\text{Shares to scale out per level} = 200/5 \text{ or } 40 \text{ (assuming an equal distribution of upsells sharesizes).}$$

The direct scale out profit to apex is \$1,200 (see Figure 11.2).

Range Zoning Method 2: The Unbounded Macrosimo Trade

Using the example above, assume that you now want to execute upbuys from \$12 to \$20, at the same \$2 intervals. These would be order entry unbounded trades. To find out the maximum shares buyable at each successively higher upbuy, we perform the following calculation:

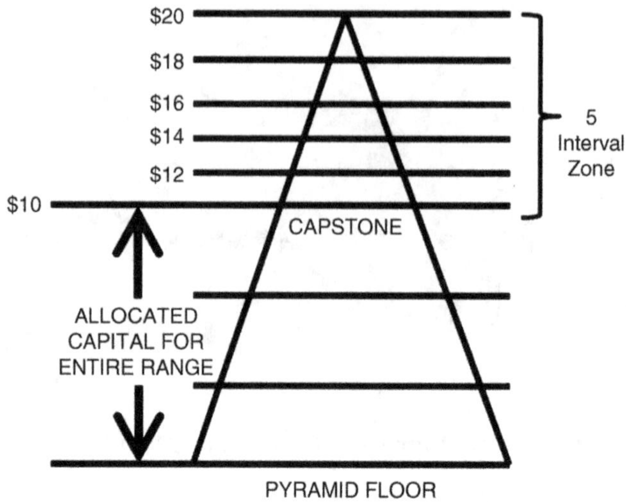

FIGURE 11.2 Range Zoning across Multiple Positions

Let x be the unknown sharesize.

$$\text{Capstone IMV} = \$2,000 = (12x + 14x + 16x + 18x + 20x)$$

$$\text{Therefore, } x = \$2,000/80 = 25 \text{ shares per level.}$$

This is purely a trend trade. Profit is determined upon being stopped out.

CHAPTER 12

Putting It All Together, Finally!

In this chapter, we shall consolidate all of the trading mechanisms and constructs into easily implementable trading plans. It would be particularly beneficial to first study the characteristics of various basic scaling constructs in terms of their performance, functionality, and interrelationship. We will briefly examine how the performance of the micro and macro vibrational returns vary with scale factoring, PLR, number of intervals, and share sizing. You will then be introduced to bounded as well as unbounded mechanisms and constructs, both vibrational and bidirectional in nature, whereupon you will learn to design pure bidirectional grids. Most importantly, always remember that all constructs must be capital bounded, but not all need be order entry bounded. This represents the minimum degree of boundedness to continue extracting vibrational and directional profits indefinitely. Capital boundedness means that the vibratrader will have sufficient, disposable, and timely capital to fund all current open longs, including all pre-planned downside constructs. Additionally, the methodology must also satisfy the Initial Entry Rule, which states that if the final account's equity is less than, or has the potential to be less than, the initial equity when price returns to the initial entry level, then the methodology is said to be unbounded.

Let us now examine the characteristics of vibrational returns under different scaling construct parameters.

THE RETURN CHARACTERISTICS OF VIBRATIONAL CONSTRUCTS

We first need to understand the table keys.

Work Cap B/E Os (D) represents the number of oscillations required to remove total used working capital, based on the downbuy price level. This would render the entire construct and pyramidal structure risk free, leaving vibratraders, if they so choose, to exit the entire ETF or commodity-based CFD with profit. Price need not return to the capstone entry level.

Run Cap B/E Os (D) is the number of oscillations required to remove total running capital based on the downbuy price level and make the entire construct and pyramidal structure break even, upon which the vibratrader may exit the entire ETF or commodity based CFD without any loss of capital. Price need not return to the capstone entry level for a break-even exit.

Micro Ret(Lo) is the percentage return over the total running capital if price vibrates at the current downbuy price level.

Macro Ret(Lo) is the maximum percentage return over total working capital if price retraces to the capstone entry level directly without additional oscillational returns.

Cum Work Cap is the cumulative working capital at the current downbuy price level.

These terms will be used in tables of various scaling parameters and their effect on the return characteristics of the vibrational construct.

A BRIEF GUIDE TO UNDERSTANDING THE SCALE ANALYSIS TABLES

Refer to Table 12.1 on page 163.

Number of Shares Purchased

The highest price level at the top of every table represents the capstone entry level. In Table 12.1, the capstone entry level is $100. The number of shares bought at the $99 downbuy level is 100, while the total number of shares bought is 200. This value is derived by adding up the shares purchased at and above the current price level of $99.

Working Cap & Cum Work Cap

We see that the working capital at that level (100 shares × \$99) = \$9,900, and that the cumulative working capital used is \$10,000 + \$9,900 = \$19,900.

Work Cap B/E OS

The number of oscillations required at the \$99 downbuy level to reduce the working capital to zero is \$19,900/100 shares, or 199 interval oscillations.

Run Cap B/E OS

The number of oscillations needed to remove all running capital (costs) is \$100/100 shares, or 1 interval oscillation.

Micro Ret (LO)

The percentage return per oscillation at the \$99 downbuy level with respect to running capital is equal to the single interval return/total running capital. \$100/\$200 × 100% = 0.5%.

Macro Ret (LO)

The percentage return per oscillation back to the entry level at \$100 from the \$99 downbuy level with respect to working capital = Full Macrosiso Return/Total Working Capital = \$100/\$19,900 × 100% = 0.5%.

TABLE 12.1 Non-Martingaling at Low PLFs

MACRO/MICRO BASED VIBRATIONAL RETURNS								
SF = 1 (FIXED SHARES PER LEVEL) 10 INTERNALS PLF = 0.09 (LOW)								
SF = 1	\$PRICE 10 INTERVAL	SHARE MULTIPLES	PER LEVEL COST	CUM LEVEL COST	WORK CAP B/E O S (D)	RUN COST B/E O S (D)	MICRO RET (LO)	MACRO RET (LO)
1	\$100	100	\$10000	\$10000	N/A	N/A	N/A	N/A
1	\$99	100	\$9900	\$19900	199	1	0.5%	0.5%
1	\$98	100	\$9800	\$29700	297	3	0.67%	1%
1	\$97	100	\$9700	\$39400	394	6	0.76%	1.5%
1	\$96	100	\$9600	\$49000	490	10	0.8%	2%
1	\$95	100	\$9500	\$58500	585	15	0.85%	2.56%
1	\$94	100	\$9400	\$67900	679	21	0.88%	3.1%
1	\$93	100	\$9300	\$77200	772	28	0.9%	3.6%
1	\$92	100	\$9200	\$86400	864	34	0.92%	4.1%
1	\$91	100	\$9100	\$95500	955	43	0.94%	4.7%

TABLE 12.2 Non-Martingaling at High PLFs

MACRO/MICRO BASED VIBRATIONAL RETURNS								
SF = 1 (FIXED SHARES PER LEVEL) 10 INTERNALS PLF = 9 (HIGH)								
SF = 1	$PRICE 10 INTERVAL	SHARE MULTIPLES	PER LEVEL COST	CUM LEVEL COST	WORK CAP B/E O S (D)	RUN COST B/E O S (D)	MICRO RET (LO)	MACRO RET (LO)
1	$10	100	$1000	$1000	N/A	N/A	N/A	N/A
1	$9	100	$900	$1900	19	1	5.2%	5.2%
1	$8	100	$800	$2700	27	3	7.4%	11.1%
1	$7	100	$700	$3400	34	6	8.8	17.6%
1	$6	100	$600	$4000	40	10	10%	25%
1	$5	100	$500	$4500	45	15	11.1%	33.3%
1	$4	100	$400	$4900	49	21	12.2%	42.8%
1	$3	100	$300	$5200	52	28	13.5%	53.8%
1	$2	100	$200	$5400	54	34	14.8%	66.6%
1	$1	100	$100	$5500	55	43	16.4%	82%

TABLE 12.3 Martingaling at High PLFs

MACRO/MICRO BASED VIBRATIONAL RETURNS								
SF = 2 (DOUBLING PER LEVEL) 10 INTERVALS PLR = (HIGH)								
SF = 2	$PRICE 10 INTERVAL	SHARE MULTIPLES	PER LEVEL COST	CUM LEVEL COST	WORK CAP B/E OS (D)	RUN COST B/E O S (D)	MICRO RET (LO)	MACRO RET (LO)
2	$10	100	$1000	$1000	N/A	N/A	N/A	N/A
2	$9	200	$1800	$2800	14	0.5	7.1%	7.1%
2	$8	400	$3200	$6000	15	1	10%	16.6%
2	$7	800	$5600	$11600	14.5	1.37	12%	29%
2	$6	1600	$9600	$21200	13.25	1.62	14%	46%
2	$5	3200	$16000	$37200	11.6	1.78	16.6%	69%
2	$4	6400	$25600	$62800	9.8	1.87	20%	102%
2	$3	12800	$38400	$101200	7.9	1.92	25%	152%
2	$2	25600	$51200	$152400	5.9	1.96	33.4%	235%
2	$1	51200	$51200	$203600	3.9	1.97	50%	402%

TABLE 12.4 Martingaling at Low PLFs

MACRO/MICRO BASED VIBRATIONAL RETURNS								
SF = 2 (DOUBLING PER LEVEL)								
10 INTERVALS PLR = (HIGH) SHARE MULTIPLES = x 1/100								
SF = 2	$PRICE 10 INTERVAL	SHARE MULTIPLES	PER LEVEL COST	CUM LEVEL COST	WORK CAP B/E OS (D)	RUN COST B/E OS (D)	MICRO RET (LO)	MACRO RET (LO)
2	$10	1	$10	$10	N/A	N/A	N/A	N/A
2	$9	2	$18	$28	14	0.5	7.1%	7.1%
2	$8	4	$32	$60	15	1	10%	16.6%
2	$7	8	$56	$116	14.5	1.37	12%	29%
2	$6	16	$96	$212	13.25	1.62	14%	46%
2	$5	32	$160	$372	11.6	1.78	16.6%	69%
2	$4	64	$256	$628	9.8	1.87	20%	102%
2	$3	128	$384	$1012	7.9	1.92	25%	152%
2	$2	256	$512	$1524	5.9	1.96	33.4%	235%
2	$1	512	$512	$2036	3.9	1.97	50%	402%

TABLE 12.5 Five Intervals without Martingaling

MACRO/MICRO BASED VIBRATIONAL RETURNS								
SF = 1 (FIXED SHARES PER LEVEL) 5 INTERVALS								
SF = 1	$PRICE 10 INTERVAL	SHARE MULTIPLES	PER LEVEL COST	CUM LEVEL COST	WORK CAP B/E OS (U)	RUN COST B/E OS (U)	MICRO RET (LO)	MACRO RET (LO)
1	$10	100	$1000	$1000	N/A	N/A	N/A	N/A
1	$8	100	$800	$1800	9	1	11.1%	11.1%
1	$6	100	$600	$2400	12	3	16.6%	25%
1	$4	100	$400	$2800	14	6	21.4%	42.8%
1	$2	100	$200	$3000	15	10	26.6%	66.6%

	5 INTERVALS		10 INTERVALS
TOTAL COST	LOWER	√	HIGHER
MICRO RET (LO)	HIGHER	√	LOWER
MACRO RET (LO)	NO CHANGE		NO CHANGE
WORK CAP B/E	LOWER	√	HIGHER
RUN CAP B/E	LOWER	√	HIGHER

FIGURE 12.1 Horizontal Analysis

Horizontal Analysis

We shall now do a simple horizontal analysis of varying numbers of intervals over a fixed SF, price range, PLR, and share multiple. Horizontal analysis means studying the effects of a fixed PLR across different intervals. From Tables 12.1 and 12.5, we draw our conclusion, as laid out in Figure 12.1.

Decreasing the intervals over the same price range and SF increases the micro-based returns, with no change in the macro-based returns. It also lowers the break-even oscillations and total invested capital.

Vertical Analysis

Vertical analysis means examining the various vibrational parameters across different PLRs. We will now do a simple vertical analysis over a fixed SF, price range, intervals, and share multiple. From Tables 12.1 and 12.2, we draw our conclusion, as laid out in Figure 12.2.

Increasing the PLR over the same price range and SF increases micro-based returns with no change in the oscillations required to recoup capital. It also lowers the total invested capital.

	FOR SAME SF = 1 AND PRICE RANGE ($9)		
	HIGH PLR		LOW PLR
TOTAL COST	LOWER	√	HIGHER
MICRO RET (LO)	HIGHER	√	LOWER
MACRO RET (LO)	HIGHER	√	LOWER
WORK CAP B/E	LOWER	√	HIGHER
RUN CAP B/E	NO CHANGE		NO CHANGE

FIGURE 12.2 Vertical Analysis

	LOW SF		HIGH SF	
TOTAL COST	LOWER	√	HIGHER	
MICRO RET (LO)	LOWER		HIGHER	√
MACRO RET (LO)	LOWER		HIGHER	√
WORK CAP B/E	HIGHER		LOWER	√
RUN CAP B/E	HIGHER		LOWER	√

FIGURE 12.3 Scale Factor Analysis

Scale Factor Analysis

Scale Factor Analysis relates to changes in the scaling of sharesizes, with the same fixed parameters. From Tables 12.2 and 12.3, we draw our conclusion, as laid out in Figure 12.3.

Increasing the SF over the same price range and PLR tends to increase vibrational returns, but it also increases the total invested cost.

Ramp Up/Down Analysis

Below is a simple Ramp Up/Down Analysis, over fixed SF, price range, intervals, and PLR. Ramp up/down analysis means increasing and decreasing the share multiples to see the effect on returns. From Tables 12.3 and 12.4, we draw our conclusion, as laid out in Figure 12.4.

Changing the share multiples does not affect the percentage returns or oscillations to break-even. It does, however, lower the total invested cost and actual profit received.

Profit Matching

The objective of profit matching is to match Microsiso and Macrosiso profits. We use this technique when we want to achieve the same Microsiso profits as the equivalent Macrosiso over the same price range.

	LOW MULTIPLES		HIGH MULTIPLES
TOTAL COST	LOWER	√	HIGHER
MICRO RET (LO)	NO CHANGE		NO CHANGE
MACRO RET (LO)	NO CHANGE		NO CHANGE
WORK CAP B/E	NO CHANGE		NO CHANGE
RUN CAP B/E	NO CHANGE		NO CHANGE

FIGURE 12.4 Ramp Up/Down Analysis

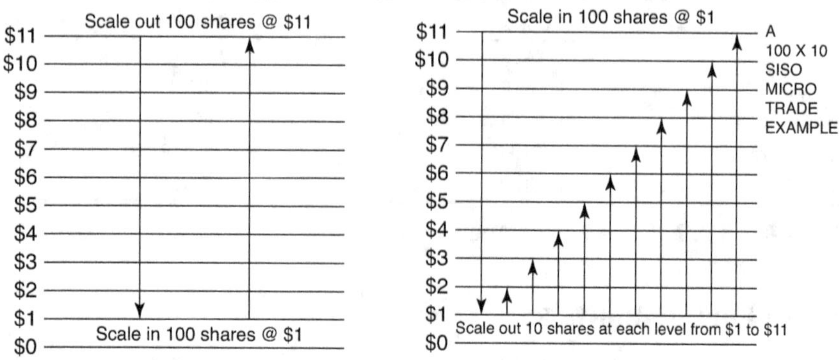

A 100 Share **Macrosiso** swing will give a profit of = 100 shares × ($11–$1) = $ 1000

A **100 × 10 Macrosiso** swing will give a profit of = $(1+2+...+10) × 10 shares = $550

FIGURE 12.5 Profit Matching

Microsiso trades will always profit less than their Macrosiso counterparts in a simple full swing, because Microsiso scales out earlier.

Refer to Figure 12.5 for an explanation of profit matching.

First, find the Share Multiplier.

Share Multiplier = Macrosiso Profit/Microsiso Profit = 1,000/550 = 1.818.

Then multiply each Microsiso share by this multiplier.

Example: 10 shares × 1.818 = 18 shares.

Therefore, in order to match up the profit derived in the Microsiso with that of the Macrosiso, we are required to buy a total of (100 × 18) 180 shares on the Macrosiso initial downbuy. We then scale out 18 shares per Microsiso upsell.

Profit matching is an extremely useful tool, especially when a vibrational construct is required to match another's profit potential in a structurally driven, negatively correlated market.

Calculating Allocated Capital, Interval Size, Scaling Factor, and Effective Price Range

In this section we will run through a simple example illustrating how to allocate capital in a construct. You will see examples of how Macrosiso and Microsiso may be traded at the same time, one for short term profit extraction and the other for a longer term trend-based return.

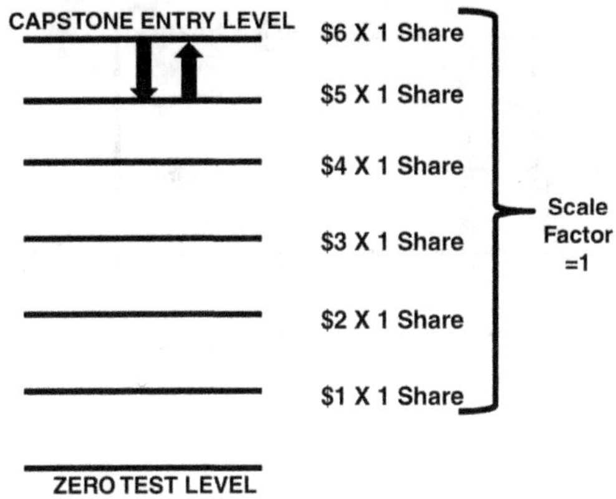

FIGURE 12.6 Construct with a Scale Factor of One

Let us assume that we have a basic vibrational construct with the following parameters:

Number of Levels = 7 Number of Intervals = 6 Interval Size = \$1
Scale Factor = 1 Capstone Entry Level = \$6

See Figure 12.6.

If price did one oscillation between the capstone level and the \$5 level, in the process upselling at the \$6 level, the vibrational return would be:

$$\text{Downbuy Sharesize} \times \text{Interval Size} = 1 \times \$1.$$

Our vibrational return would be \$1.

But what would returns be if the scale factor was 2?

$$\text{Downbuy Sharesize} \times \text{Interval Size} = 2 \times \$1 = \$2.$$

For the next question, refer to Figure 12.7.

What vibrational returns would be generated if price oscillated from the capstone level to \$3 and back if we used the Microsiso mechanism?

The vibrational returns would be \$3, since each upsell generated \$1.

If the scale factor was 5, the vibrational returns would be equal to (\$5 + \$25 + \$125) = \$155.

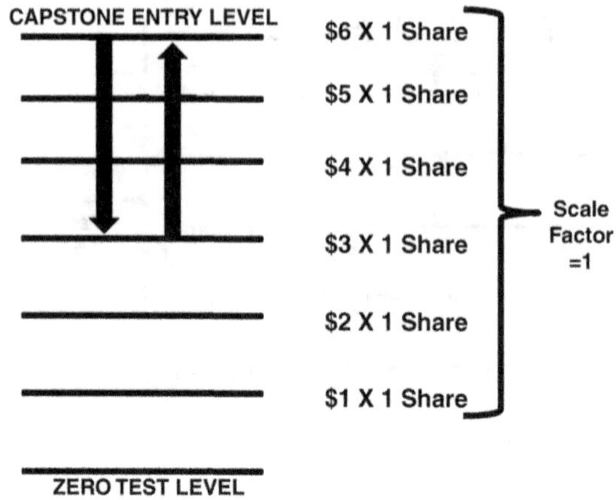

FIGURE 12.7 Extracting Returns via the Microsiso

The capstone level starts off with 1 share, followed by the $5 level with 5 shares, the $4 level with 25 shares, and finally the $3 level with 125 shares. Each level's sharesize is multiplied by a factor of 5 to get the next downbuy sharesize.

To determine the total allocated capital, running capital, and working capital at a scale factor of 2 if prices fell to and remained at the $2 level, see Figure 12.8.

$$\text{Total allocated capital} = (\$6 \times 1) + (\$5 \times 2) + (\$4 \times 4) + (\$3 \times 8) + (\$2 \times 16) + (\$1 \times 32) = \$120.$$

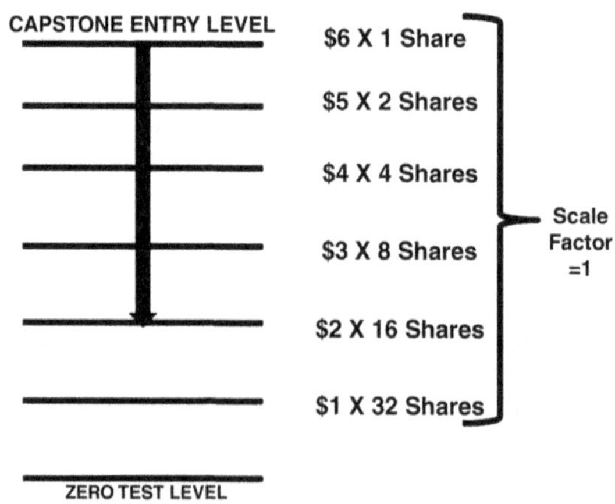

FIGURE 12.8 Construct with a Scale Factor of Two

We would only need all this if price tests the \$1 level. It represents the total capital required to fund all the downbuys in the construct.

$$\text{Working capital at the \$2 level} = (\$6 \times 1) + (\$5 \times 2) + (\$4 \times 4) + (\$3 \times 8) + (\$2 \times 16) = \$88.$$

Working capital is the capital required to fund all downbuys to the \$2 level.

$$\text{Running capital} = (1 \times \$4) + (2 \times \$3) + (4 \times \$2) + (8 \times \$1) = \$26.$$

Running capital represents the amount of decline in the total value of the shares purchased. It is the difference between the IMV and CMV. It can also be expressed as the difference between CVAR and CMV.

Now, what would our returns be if we used the Macrosiso mechanism to upsell all the downbuys at the capstone level, from the \$2 level upwards?

$$\text{Returns} = (16 \times \$4) + (8 \times \$3) + (4 \times \$2) + (2 \times \$1) = \$98.$$

Here 16 shares are exited across \$4, the next 8 shares are exited across \$3, the next 4 shares are exited across \$2, and the last 2 shares are exited across \$1.

Using the same example, how many oscillations would price have to make at the \$1 level in order to recoup or remove the working, running, and total allocated capital at the \$2 level?

At the \$1 to \$2 level, each oscillation will yield a vibrational return of \$32, 32 shares × \$1.

Therefore, the number of oscillations to remove working capital = \$88/\$32 = 2.75.

The number of oscillations to remove the running capital = \$26/\$32, less than one.

The number of oscillations to remove total allocated capital = \$120/\$32 = 3.75.

This means that it would take under 3 oscillations to make your open longs risk free, while it would take under 4 oscillations to make the entire vibrational construct risk free!

As another example, what returns would we make if we exited half the shares at the capstone level, then exited the rest at each next higher upsell from the \$2 level? This means that we would trade the Microsiso and Macrosiso in tandem.

The total Micro and Macrosiso trade yields \$49 + (\$8 + \$4 + \$2 + \$1) = \$64.

Finally, if we had an ETF that oscillated across a wide range of prices, but never above the \$12 level nor below the \$5 level, what would the effective range of the vibrational construct be, bearing in mind that we have only enough capital to place our capstone entry level at \$9?

The effective range is usually measured from the historically highest price in an ETF, but will be superseded under the condition of capital insufficiency. So in this case, the effective range will be \$9.

INTRODUCTION TO VIBRADIRECTIONAL TECHNIQUES

In this section, we shall learn to combine vibrational trading with trend capture or directional techniques, giving rise to what I call Vibradirectional Trading. Vibradirectional trading combines the best vibrational and directional mechanisms to form the basis of a trend-vibrational construct. We will incorporate various bounded and unbounded mechanisms covered earlier into new trend and vibration capture techniques.

It is imperative to understand that the basis for all vibradirectional and vibra*bi*directional trading is capital boundedness. We only allow unboundedness in the form of order entry unboundedness. By this we mean that we will only use stop orders for:

1. Exiting or locking in a profit
2. Creating a breakeven entry
3. Initiating entries.

Also, we must make sure that we are able to fund long positions down to zero test level, except when longs are in a break-even or risk free state. In that case, positions are exited by a stop order placed at the entry price, thereby freeing up the capital in a low margin account for additional long entries. Remember, all entries must be initiated from either the apex or the capstone entry level. This means you enter at a price level from which you would be able to fund all upbuys and downbuys from the floor to the apex. If trading commodity CFDs on margin, then you must be able to hold all open long positions down to zero test level including maintenance margin, unless the positions are rendered break-even or profit has been locked in via a stop order.

Using the Bounded Macrososi for Range-bound Vibrational Profit Capture

We have seen in earlier chapters how the bounded Macrososi works to extract vibrational returns. Since we initiate the Macrososi at the capstone entry level, this means that should price decline, we would be able to convert our Macrososi into either the Macrosiso or Microsiso for generating profit within the pyramidal base. Should price rise above the capstone entry level, we could then revert back to Macrososi trading seamlessly. That is the power and beauty of trading from the capstone entry level.

The problem with vibratrading the Macrososi is that all trading terminates at the apex, since all longs would have already been scaled out by that point. Therefore, to access price levels beyond the apex we can execute a buy and hold at any point from the apex to the capstone entry level. This will give us a chance to participate in any upside move beyond the apex for additional profit. Note that our position is now order entry unbounded as we have used a stop loss to lock in trend profit. If price should gap back into the pyramidal structure below the capstone entry level, we would not exit the original Macrososi long positions with a loss since the longs were entered within the pyramidal structure. If we bought additional longs beyond the apex, then our positions would be unbounded, as we would suffer a loss exit should price gap below the original entry level. The problem here is that the new longs were not initiated within the pyramidal structure, so there would not be sufficient disposable funds to hold all positions till zero test level.

Unbounded Upside Trend and Vibrational Profit Capture Techniques

Before we introduce vibradirectional and vibrabidirectional techniques, we need to look at one basic mechanism. Dual sizing is simply the doubling of a position or share size. By dual sizing, we are able to experience both

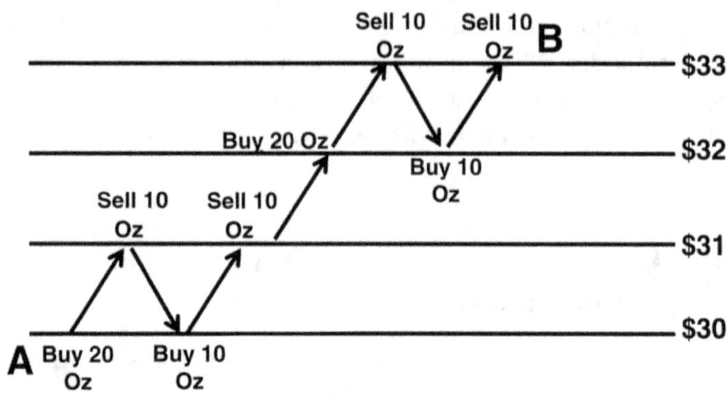

FIGURE 12.9 Extracting Upside Vibrational and Trend Profit

trend and vibrational returns without sacrificing one for the other. Of course, we may also vary the ratio of trend to vibrational exposure by allocating a larger position or share size to the trend or vibrational positions via multi-sizing.

Let us examine the mechanics of dual sizing. We see straightaway in Figure 12.9 that if we bought 10 troy ounces of silver via a CFD at \$30, we would be able to exit this position at the next upsell at \$31 for a profit of \$10. If price should retest \$30, we re-enter the same long silver position and repeat the upsell for another \$10 of vibrational profit. As long as price oscillates around the \$30 to \$31 levels, we would carry on generating a profit. But should silver climb to \$32, we would need to re-enter a new long at that level. As you noticed, we did not earn any trend profit in the process.

However, by increasing our position in silver to 20 ounces at \$30, we keep half our position and make it possible to sell the remaining 10 ounces for the purposes of generating some trend profit should price rise. We then would have made approximately \$20 from the vibrational trading and an extra \$10 from trending action. (\$32–\$31) × 10 oz. = \$10.

Total Vibrational Profit from Point A to B is \$10 × 4 = \$40.

Total Trend Profit from Point A to B is \$30 + \$10 = \$40.

As we can see, dual or even multi-sizing helps the vibradirectional trader capture directional and vibrational profit simultaneously. Let us now look at dual sizing to the downside (see Figure 12.10).

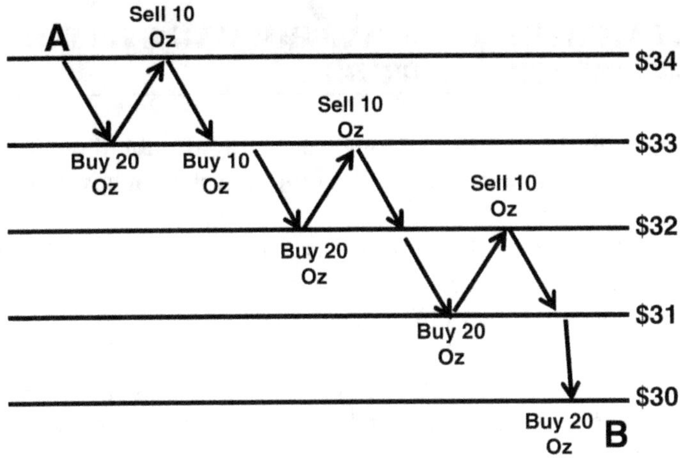

FIGURE 12.10 Extracting Downside Vibrational and Trend Profit

Here we see the dual-sized Microsiso mechanism generating oscillational returns at every upsell as price declines.

The total vibrational return extracted between Point A and B is $10 × 3, or $30.

Trend profit is derived from positions held across at least two intervals. This means that all Microsiso mechanisms are designed to extract vibrational and not trend-based returns. As we can see, there is no trend profit as of yet.

We shall see later that dual sizing affords the vibratrader an opportunity to accumulate additional risk-free positions at various price intervals. By risk free we mean that these accumulated positions are protected by a break-even stop and need not be held until the zero test level. It also implies the ability to accumulate unlimited longs without any added capital, one of the greatest advantages of dual sizing. Order entry unboundedness is still present.

To recap, dual sizing allows the vibratrader an opportunity to experience vibrational as well as trend profits, while also accumulating risk free positions without any added capital. It also allows us to reduce the probability of getting stopped out by incorporating defensive accumulation, which we will be covering shortly.

CALCULATING WORKING AND RUNNING CAPITAL WITHIN VIBRATIONAL GRIDS

Before we begin our journey through the array of vibradirectional and vibrabidirectional trading constructs, a little refresher on how to calculate the downside exposure may be useful.

We shall be using XAG (silver) as a commodity-based CFD for the rest of our examples. These calculations are similar to those used with ETFs, except for a few important differences:

1. We need to account for the margin when using CFDs to make sure that we have sufficient running capital to hold all positions down to the zero test level.
2. We actually free up all working capital in our CFD longs positions once break-even entry is achieved, except for a small margin. ETFs, on the other hand, require additional capital funding since most stocks are traded on either a fifty percent margin or none whatsoever. Break-even entries for all the lower level long positions in an ETF or stock remove the running capital, but leave a large portion of the working capital due to the higher margin requirement. As a result, many of the vibradirectional constructs will be more costly to implement in an ETF.

Note: The profit or loss per \$1 move in any precious metal price will equate to the amount of troy ounces held. Hence, exiting a 20 oz. long position at a level \$2 above the entry will equate to a profit of \$40 (\$2 × 20 ounces).

Calculating the Minimum Downside Working Capital

Assume equidistant \$2 intervals within a price range of \$30 to ZTL on a CFD platform. Let the minimum lotsize for XAG be 10 ounces and the required margin be 0.25 percent on the CFD platform. Refer to Figure 12.11.

$$\begin{aligned} \text{Average Price} &= (\text{Highest long position} + \text{lowest long position})/2 \\ &= (\$30 + \$2)/2 = \$16. \\ \text{Number of Positions} &= (\text{Range of Long Positions/Interval Size}) + 1 \\ &= ((\$30 - \$2)/\$2) + 1 = 15. \end{aligned}$$

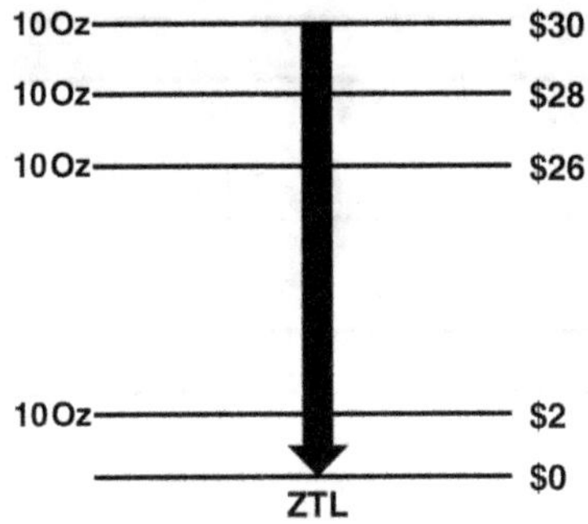

FIGURE 12.11 Finding the Downside Working Capital

Minimum Working Capital Required without Margin

$= 15 \times 16 \times 10$ oz. $= \$2{,}400$.

Minimum Working Capital Required with Margin

$= \$2{,}400 + (\$2{,}400 \times 0.0025) = \$2{,}406$.

Minimum Working Capital Required for Dual Sizing Applications with Margin

For dual sizing, we require at least double the minimum lotsize:

$$= (15 \times 16 \times 20 \text{ oz.}) + \text{Margin}$$
$$= \$4{,}800 + (0.0025 \times \$4{,}800) = \$4{,}812.$$

Please note that for a CFD vibratrader, downside working capital represents the maximum running capital. The vibratrader must therefore have these funds in the trading account plus margin, or have access to them when required. As we will see later, the CFD vibratrader has no real advantage over the ETF vibratrader to the downside. But to the upside there is a vast difference.

Non-Equidistant Interval Sizing

If the vibrational scaling construct has non-equidistant intervals, we cannot use the formula above for the number of positions. We may account for the total working capital via simple addition with respect to each position, as indicated in the example in Figure 12.12.

Next we will calculate the total working capital of all positions.

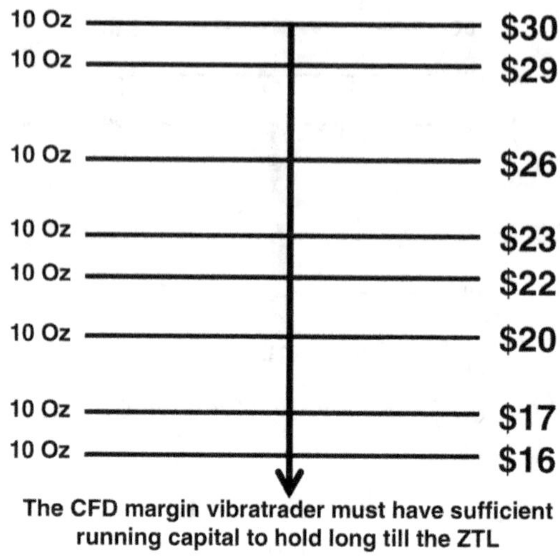

FIGURE 12.12 Non-Equidistant Interval Sizing

Assume a minimum lotsize of 10 oz. per level for all positions.

$$\begin{aligned}\text{Total working capital} &= (\$30 + 29 + 26 + 23 + 22 + 20 + 17 + 16) \\ &\quad \times \text{Lotsize} = 183 \times 10 \\ \text{Total working capital} &= \$1{,}830.\end{aligned}$$

On the other hand, if we wanted to calculate the maximum lot size per level for a given working capital, we would simply invert the formula, assuming equal lotsizing per level and a working capital of \$2,800.

$$\text{Lotsize per level} = \text{Working capital}/183 = \$2{,}800/183 = 15.30 \text{ ounces.}$$

Concerning Lotsize Increment on the CFD Platform

There is a minimum lotsize requirement on all platforms. There will be many times when the CFD vibratrader cannot allocate the calculated lotsize due to the minimum lotsize increments for entries. For XAG, CFD providers generally offer a range of minimum lotsize increments from 1, 10, 50, or upto 500 ounces.

The vibratrader should trade at a lower lotsize until the size of the position satisfies the next higher lotsize increment. Generally, it is preferable to undertrade rather than overtrade.

Shortcut for Calculating the Running Capital

Assuming that the intervals are equidistant and the lotsizes are equal, we may use a simple mathematical shortcut for calculating the running capital.

To calculate the sum of a series of consecutive numbers, we simply use the following formula where N = range of numbers:

$$N\,(N+1)/2$$

Example: Sum of 1 to 10 is $10 \times (10 + 1)/2 = 55$.

This is a very useful formula for calculating the sum of cumulative intervals within a construct. Cumulative intervals are the total intervals traversed by all the long or short positions within the construct.

Calculate the Running Capital at Equidistant Intervals to a Specific Level

Assume an equal lotsize per interval of 10 oz. with an interval size of \$2 (see Figure 12.13).

Number of intervals = (Range/Intervals) = (\$30 − \$16)/\$2 = 7.
Number of cumulative intervals = N(N + 1)/2 = (7 × 8)/2 = 28.
Running capital = Cumulative intervals × Lotsize × Intervals size
= 28 × 10 oz. × \$2 = \$560.

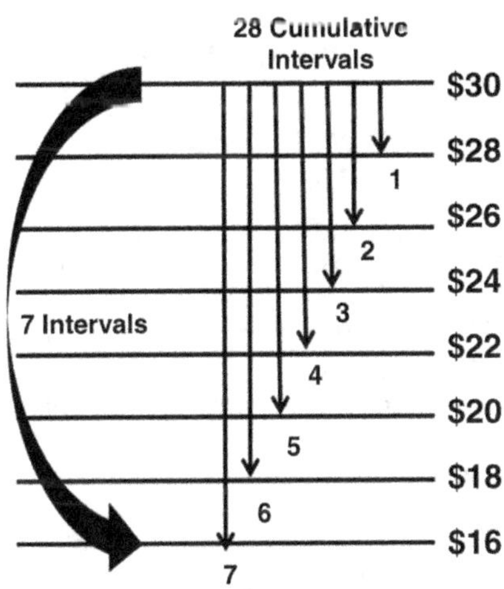

FIGURE 12.13 Cumulative Intervals

Therefore the vibratrader will require at least \$560 (plus some margin) in the trading account in order to hold all the long positions down from \$30 to \$16.

The Relationship between Working Capital, Running Capital, and CMV

See Figures 12.14 and 12.15. As seen earlier:

$$\begin{aligned}\text{Current Market Value (CMV)} &= \text{Total lotsize} \times \text{Current price}\\ &= 70 \text{ oz.} \times \$24 = \$1{,}680.\\ \text{Working capital} &= \text{CMV} + \text{Running capital}\\ &= \$1{,}680 + \$140 = \$1{,}820.\end{aligned}$$

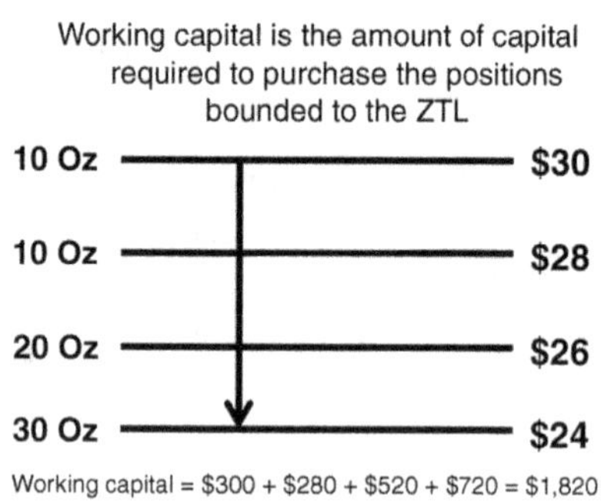

FIGURE 12.14 Finding the Working Capital

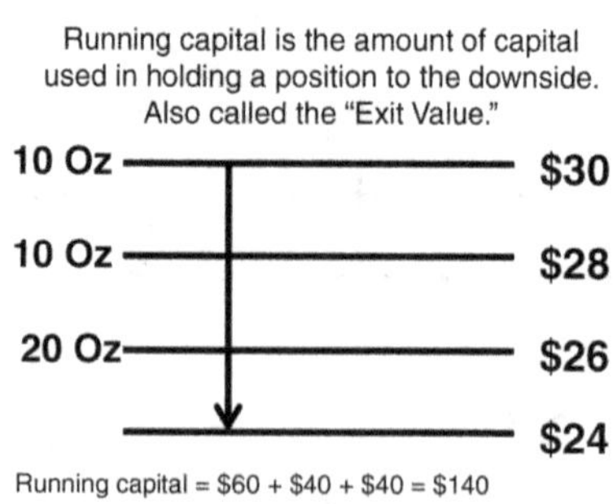

FIGURE 12.15 Finding the Running Capital

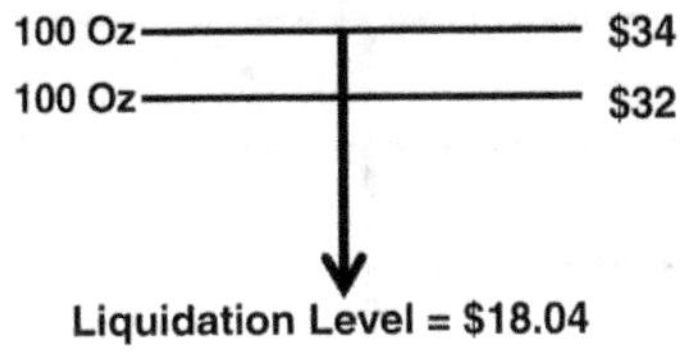

FIGURE 12.16 Finding the Liquidation Level

Calculating ZTL and Liquidation Levels within Vibrational Grids

Let us assume that we have only $3,000 of working capital. At what level would we experience the forced liquidation of our long positions? Assume the cut off margin percentage level is 100 percent, lotsize per level is 100 ounces, and account margin is 0.25 percent (see Figure 12.16).

$$\begin{aligned} \text{Average price} &= (\$34 + \$32)/2 = \$33. \\ \text{Number of positions} &= (\text{Range/Intervals}) + 1 = ((\$34 - \$32)/\$2) + 1 = 2. \end{aligned}$$

Using a simple formula:

$$(\text{Average price} - \text{Liquidation price}) \times (\text{Number of positions x Lotsize per level}) = \text{Capital margin}$$

Rearranging, we get:

$$\begin{aligned} \text{Liquidation price} &= \text{Average price} - ((\text{Capital} - \text{Margin})/ (\text{Number of positions} \times \text{Lotsize per level})) \\ &= \$33 - ((\$3{,}000 - (\$3{,}000 \times 0.0025)/200) \\ &= \$18.04 \end{aligned}$$

This means that once price retraces to $18.04, the vibratrader will not have any funds to sustain the 200 ounce positions, which may then be liquidated by the broker.

Finding the Maximum Bounded Lotsize

The following example will show you how to calculate the largest lotsize allowable at a ZTL level which is higher than zero. Should price fall below this higher ZTL, the vibratrader would not be able to maintain the long

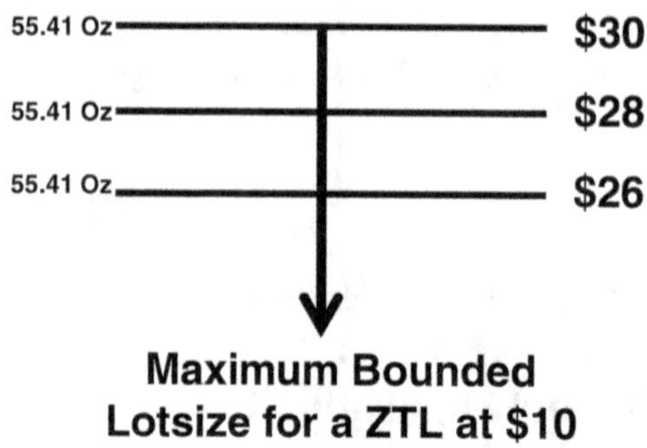

FIGURE 12.17 Position Sizing for ZTL above Zero

positions. Also, should the lotsize be larger than the calculated value, then the ZTL level would be higher than the one used. Capital is only bounded until the ZTL, no matter which price level it may be (see Figure 12.17).

Assume new ZTL = \$10, working capital = \$3,000, and margin = 0.25 percent.

Average price = (\$30 + \$26)/2 = \$28.
Number of positions = (Range/Intervals) + 1 = ((\$30 − \$26)/\$2) + 1 = 3.
Lotsize per level = Working capital/((Average price − New ZTL) × Number of positions)
Lotsize = Working capital/(\$18 × 3) = \$3,000/54 = 55.55 ounces.
Accounting for margin, lotsize = 55.55 − (55.55 × 0.0025) = 55.41 ounces.

This means that the vibratrader will be able to hold the long positions down until the new ZTL at \$10 as long as the lotsize does not exceed 55.41 ounces.

Using the Balance Curve to Monitor Boundedness

The balance curve only records the P/L of closed positions. Hence it will always be uptrending and never downtrending, as downtrending indicates that a position has closed in loss. This means that the vibrational construct is now capital unbounded. To check whether your vibratrading is bounded or otherwise, just consult the balance curve as shown in Figure 12.18.

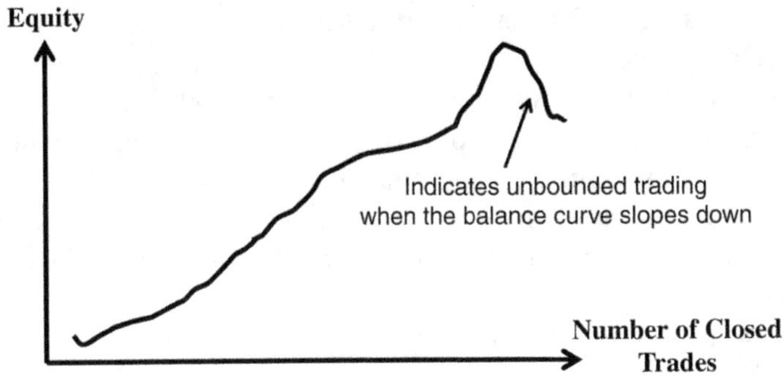

FIGURE 12.18 Balance Curve

The four main reasons the balance curve slopes downward are:

1. Positions being exited with a loss
2. Negative slippage via the use of stop orders
3. Negative swap or rollover interest
4. Using up the premature profit in a Type 2 exit mechanism before the stop is rolled up to the original entry level.

FREE SWING WITH CONSTANT CAPITAL PER LEVEL WITH TYPE 1 (ROLL TO BREAK-EVEN)

The first upside trend capture construct is called Free Swinging. This has proven an extremely popular technique as it produces the greatest returns for the same amount of working capital. The method uses what I call a Type I Exit, which is basically just a break-even stop.

Let us assume that we have an investor and a vibradirectional trader. They are both extremely keen to participate in the potential upside in silver, but each have only $3,000 working capital.

The investor directly proceeds to the nearest silver bullion dealer and purchases $3,000 worth of silver at $30 per troy oz. (We shall disregard the premium associated with buying bullion for the sake of convenience.) This gives the investor approximately: $3,000/$30 = 100 troy ounces.

The vibradirectional trader, on the other hand, funds a CFD account with an equal working capital of $3,000 and enters a long position at $30 for

100 ounces of silver. As far as we can see, neither party has any significant advantage at this point.

Let us assume that silver now falls below the $30 entry level. Neither the investor nor the vibradirectional trader is able to initiate any new long positions as all the working capital has been allocated to the 100 ounce position at $30. All they can do is wait for price to rebound and penetrate the $30 level to the upside. Until then, they are both essentially in a buy and hold scenario.

If price rises above the $30 entry level, however, the return characteristics differ significantly. The investor will make a steady $100 per $1 increase in silver price. There is no rate of increase in returns. This is where the vibradirectional trader outperforms the investor.

The Free Swing Trend Capture Technique

The vibradirectional trader above allows price to oscillate freely around the $30 entry level until it rises to a predetermined level whereby the trader will enter a new position. Let us assume that the new entry level is $31. At $31 the vibradirectional trader proceeds to create a breakeven long position by placing a stop loss order at $30. The trader could also place the stop loss order at $30.10, earning a small return should price hit the stop.

The entire working capital, save the very insignificant margin, is now freed up! The vibradirectional trader may now place a new lot at $31 with the same $3,000 working capital.

$$\text{The new lot size is } \$3{,}000/\$31 = 97 \text{ ounces, rounded up.}$$

Capital boundedness is automatically preserved. The vibradirectional trader cannot allocate any lotsize larger than what the working capital can fund at that new level. See Figure 12.19.

The vibradirectional trader frees up each lower level before initiating a new long position. This is crucial because, should the trader be caught in a rapid price decline before removing risk from the lower levels, that trader will be holding on to an unbounded construct.

If that rule is not violated, however, we see from the example above that the vibradirectional trader is actually exposed to tremendous upside returns, holding nearly 720 troy ounces at the $37 price level.

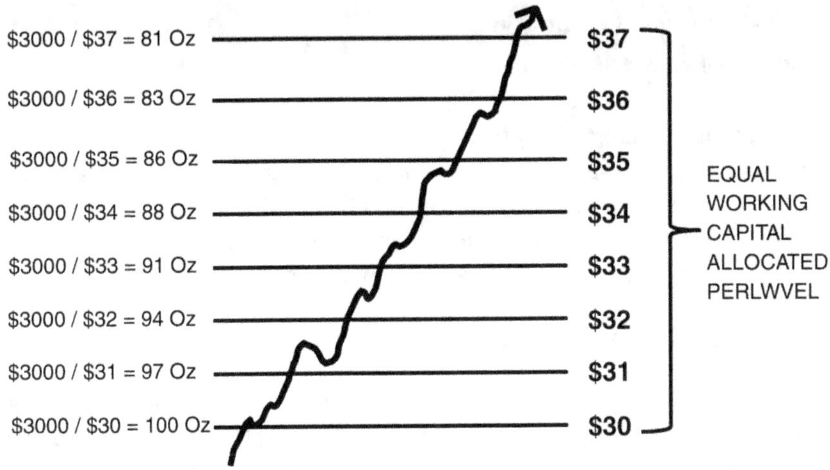

FIGURE 12.19 Equal Capital Allocation per Level

The running P/L will be:

$$= (7\times\$100)+(6\times\$97)+(5\times\$94)+(4\times\$91)+(3\times\$88)+(2\times\$86)+(1\times\$83)$$
$$= \$2,635.$$

The vibradirectional trader experiences a return of $2,635 at $37. This is a percentage return of 87.8 percent. The investor, on the other hand, has made a profit of only 100 oz. × ($37 – $30), which comes to $700, or a 23 percent return on that same working capital. Neither the investor nor the vibradirectional trader invested more working capital at any point. Therein lies the phenomenal advantage of vibradirectional trading.

Exponential Returns

We also see that the free swing trend capture technique affords the vibradirectional trader exponential returns of:

- $100 between $30 and $31
- $197 between $31 and $32
- $291 between $33 and $33
- $382 between $33 and $34
- $470 between $34 and $35

- $556 between $35 and $36
- $639 between $36 and $37.

If we plot a chart of return per $1 increase in price against price increment, it will assume the following curve seen in Figure 12.20.

As we can see, though the rate of return is increasing, this rate of increase is slowly decelerating due to the reducing lotsize at each higher level. This is because we have kept our working capital constant.

The Self-Financing Effect

As price rises, the lotsize starts to decrease in order to remain bounded at constant working capital. But if we lock in profit at the lowest levels by rolling up the stops, we may be able to raise our total working capital. Therefore, we

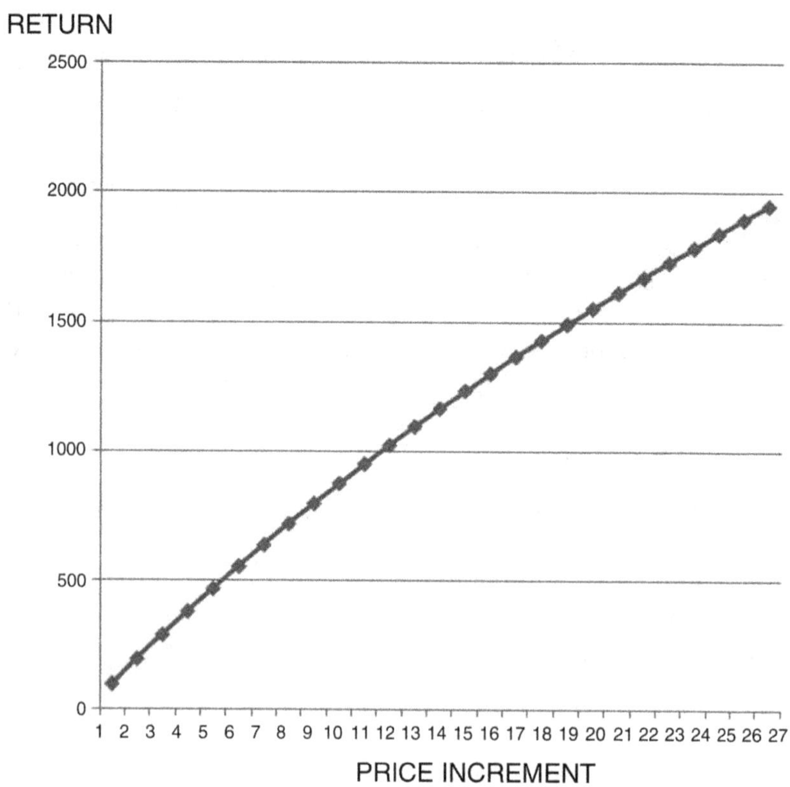

FIGURE 12.20 Exponential Returns via Free Swinging

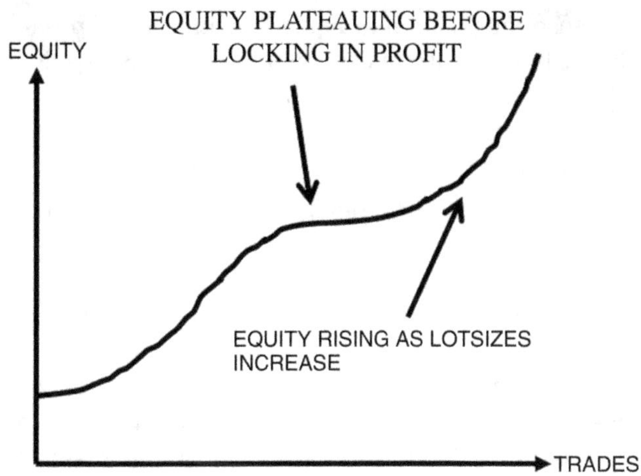

FIGURE 12.21 The Self-Financing Effect

can allocate a larger lotsize as price rises, which in turn starts to increase our equity. This is the self financing effect (see Figure 12.21).

GAPS IN THE GRIDS

One of the issues associated with free swinging is the creation of gaps in the construct when price hits the break-even stops at successively lower levels. See Figure 12.22.

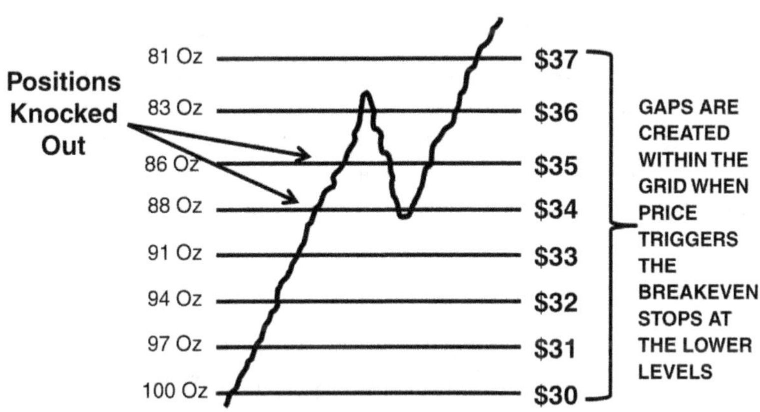

FIGURE 12.22 Gaps in the Grids

This will cause the position to liquidate with a small profit, since the stops are placed 10 cents above each entry price. This positioning of the stops helps to:

1. Cover any negative rollover interest charges
2. Reduce the effects of negative slippage associated with the use of stop orders.

Though this is one of the characteristics of the free swinging technique, it is, as they say, a good problem! The vibradirectional trader did not permanently deplete any capital in the process, and in fact still has the advantage of significantly more ounces per working capital as compared to the investor. All the vibradirectional trader has really lost is an opportunity to earn more should price rise.

To reduce gaps in the grids we may employ any of the following methods:

1. Dividing the upside capital into multiple parts
2. Use of the free styling technique
3. Use of downside capital
4. Use of the zero cost hedging technique
5. Use of the defensive accumulation technique.

Dividing the Upside Capital into Multiple Parts

By dividing our upside capital into three parts, for example, we may re-enter any gap once price starts to rebound to the upside. Of course, we need not initiate a lotsize that utilizes the full capital allocated to that part. We can always initiate a long with reduced lotsize, leaving us more capital to fill gaps should price continue to decline and take out our breakeven stops at the lower levels.

Use of the Free Styling Technique

We may also use the free styling technique to reduce the incidence of gaps being created. Free styling is simply the allocation of working capital across a number of consecutively higher levels, instead of allocating the full upside capital just to one level. We shall cover free styling further very shortly.

Use of Downside Capital

We may also use capital that was specially allocated for significant moves to the downside. This downside working capital must not be mixed with the upside allocation.

Use of the Zero Cost Hedging Technique

Zero cost hedging can also help reduce the incidence of gaps. Zero cost hedging is a technique whereby the vibradirectional trader initiates equal and opposite orders across the same price levels. The trader may then cover as many shorts as required to take advantage of declining prices, by having all long position running costs funded by the short profits. The trader must be able to fund all the exposed longs until the ZTL. If circumstances turn out to be disadvantageous, the trader simply squares off all pairs, incurring only a small cost. Zero cost hedging will be described in more detail shortly.

Use of the Defensive Accumulation Technique

The use of this technique will also reduce the occurrence of gaps within the construct. Basically, we are trying to establish a long position whereby the amount of positions exited will enable a stop to be placed at a significant distance below price entry, making it harder to stop out. We will also be covering this very shortly.

THE BALANCE BETWEEN OPPORTUNITY COST AND PROFITABILITY

The vibradirectional free swing trader may risk slip-through by making the intervals too wide. On the other hand, by making the intervals too narrow, the trader may experience a rise in opportunity cost from the increasing amount of stop outs caused by having the break-even stops too close to the next entry level. This results in the rapid increase in gaps being created by price retracing to take out these break-even stops at the lower levels (see Figure 12.23).

As we can see from the narrow interval sizing above, only the position indicated by the arrow survived being hit by price throwbacks to the previous break-even level. Therefore, it is not hard to imagine a scenario

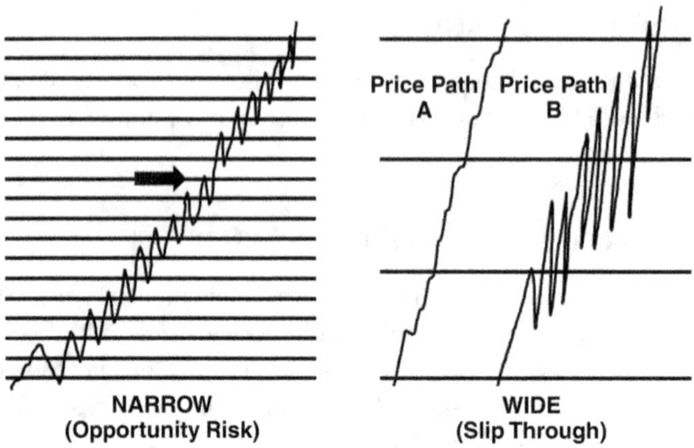

FIGURE 12.23 Narrow vs. Wide Interval Sizing

where vibradirectional traders enter successively higher positions furiously at very fine intervals only to find that they have hardly any positions left after an extended run-up in price. It should be noted that there is no issue of opportunity risk for narrow interval sizing to the downside, because there is no risk-freeing as price declines for long positions. Therefore the intervals may be as narrow as possible, but it will require more working capital to support the downside construct.

On the other hand, if the intervals are too wide, the vibradirectional trader will tend to experience more trend profit than vibrational. As seen in Figure 12.23, both price paths yielded the same return at the end of the price run-up because there were no vibrational returns generated in price path B.

Free Swinging with SLV and CFDs

As mentioned earlier, there is essentially very little difference between the use of SLV and an XAG (silver) CFD to the downside. The SLV trader must have the funds to position longs at successively lower intervals, assuming for now that there is no access to margin. The CFD vibradirectional trader will also need this amount in the CFD account to withstand price declines to near ZTL should they occur. This is because CFD vibradirectional traders are not allowed to use a stop loss mechanism, under the principles of capital boundedness.

But to the upside, the CFD vibradirectional trader must only have enough capital and margin for the number of longs that are still not rendered risk free by either a Type 1 or Type 2 break-even exit. The CFD vibradirectional trader must be able to hold the last-in position all the way down to near ZTL. The last-in position is initiated at the very highest price level which has yet to be rendered risk free by a break-even exit. The CFD vibradirectional trader is able to control up to 780 ounces of silver for an initial working capital of $3,000. This is a vast difference from the investor who only controls 100 ounces with the same working capital.

The SLV vibradirectional trader is not afforded the same advantage. Each successively higher long position must be initiated with new capital. Therefore, for the SLV vibradirectional trader to control the same number of ounces as the CFD vibradirectional trader, the SLV trader will need a total of $3,000 × 7 levels, or $21,000 (should silver run from $30 to $37 as seen in our previous example). This is 600 percent of the initial working capital of $3,000. The CFD vibradirectional trader only required the initial working capital, and nothing more.

One last issue that the SLV vibradirectional trader must endure is the limited trading hours available. SLV only trades during the stock exchange trading hours. This may lead to gapping price action on the open as price adjusts to the new levels in silver. But the CFD vibradirectional trader is able to trade price action twenty-four hours a day and thereby avoid many of these gapping issues.

Upside Scalping Technique for Free Swinging

We shall now turn our attention to a technique for extracting additional returns from minor price fluctuations between the free swing predetermined standard entry levels. The vibradirectional trader is able to generate additional scalping profit without adding capital.

You enter the construct at the usual $1 intervals from $30 upwards, as described earlier. But you may also enter smaller positions between the standard free swing entry levels. In Figure 12.24, the vibradirectional trader enters a new position at $30.30 with 50 ounces, once the $30 level has been freed of risk. The trader exits the 50 ounce position at $30.60 (Point 2) for a profit of $50 × $0.30 = $15. Once price returns to the $30.30 level (Point 3) a new 50 oz. position is opened and the process is repeated until either price hits the $31 level (Point 4) or declines below the initial entry price of $30.30 (Point 5).

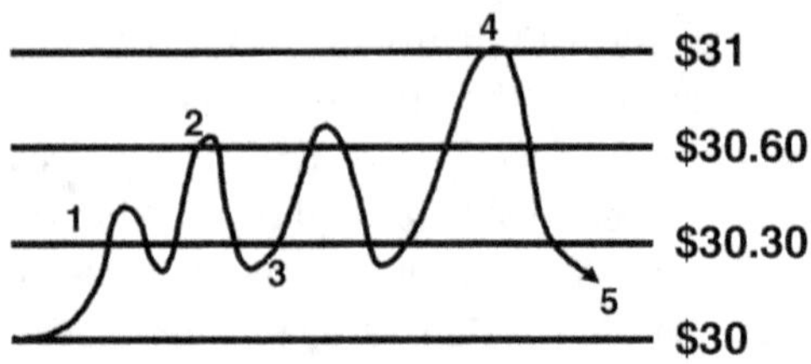

FIGURE 12.24 Upside Scalping

Once price hits the $31 level, which is now the "last-in" position, no further re-entry is permissible at the lower levels, assuming of course that the vibradirectional trader has allocated full working capital at each standard free swing entry level. As we can see, upside scalping requires no additional capital even though the vibradirectional trader has fully allocated all funds to the position at $30.

Other possible ways of exiting an upside scalp position may include:

1. Exiting a few cents below the $31 level in hopes that price retraces to the $30.60 level, thus affording the vibradirectional trader a new upside scalping entry.
2. Implementing a trailing stop at a few cents below the standard $31 free swing entry level once it is tested. This will give the vibradirectional trader an opportunity to experience a greater return should the trailing stop remain intact.

Upside scalping may transform into pure vibrational trading. This is also true for positions in which a single minimum lotsize is entered and exited across two adjacent levels, only allowing a new entry at every other higher level.

In Figure 12.25, if the vibradirectional trader enters a 10 oz. lot at $30, it may be exited at $30.50 for a profit. If price continues to the upside, the next new long entry is at $31.50, for another 10 ounces. There is no point in exiting and re-entering at the $30.50 level as that would simply equate to a buy and hold position. That is why there is no point in having a Microsiso operating to the upside—it is redundant.

Nevertheless, we see that the upside scalp, as well as the single minimum vibrational lot, may transform into pure vibrational trading only after an exit.

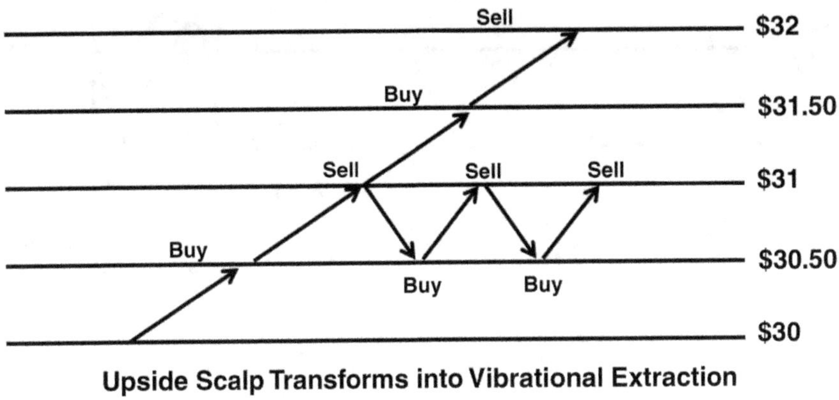

FIGURE 12.25 Single Sized Position in the Market

Free Swing with Constant Capital per Level with Non-Accumulation

In this upside trend capture technique, we initiate longs of equal lotsize as price rises. Let us start with a very basic construct (see Figure 12.26.)

No accumulation or trend profit extraction is possible with just a single minimum lot, though vibration profits are possible in a single lot construct. Let us now turn our attention to a dual sized version.

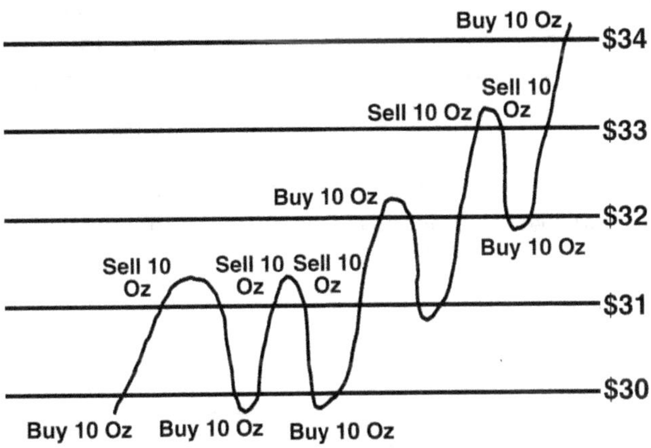

FIGURE 12.26 Free Swinging with Non-Accumulation

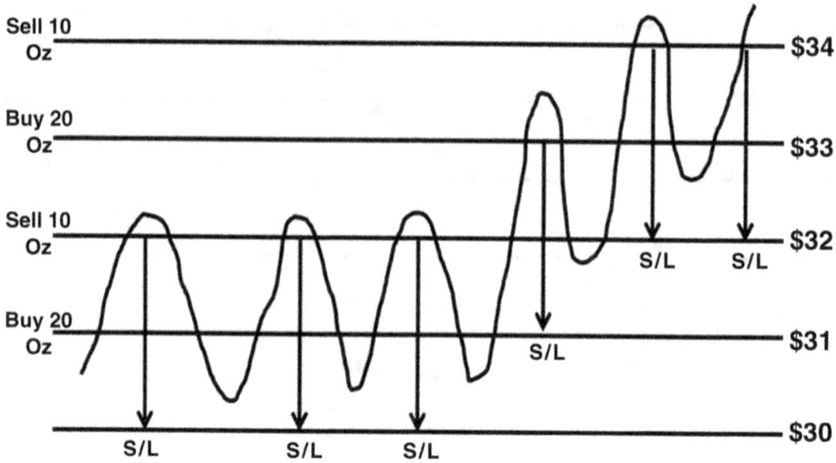

FIGURE 12.27 Free Swinging with Type 2 Exit

Free Swing and Type 2 (Scale to Break-Even) with Single Interval Accumulation

In Figure 12.27, we use a Type 2 Scale to Break-even exit. This means that we enter a 20 ounce long position at $31 and exit 10 ounces at $32. We now place a stop at $30 for the remaining 10 ounces. This will create a risk-free position at $31. After every profitable exit at $32, we re-enter the full 20 ounces at $31, since we have already freed from risk the original 20 oz. position. Therefore, no added capital is required to accumulate positions at the $31 level. Notice too that Type 2 exits make it harder for price to hit the stops, which are two intervals below the exit price. The position has less risk of being taken out by falling price action.

Finally, we see that the construct above allows us to accumulate only across certain single intervals, for example between $31 and $32 and again between $33 and $34. Accumulation is not possible between levels $32 and $33 above.

Taking Type 2 Profits

The vibradirectional trader must be very careful not to use the profits generated by a Type 2 exit prematurely. Until the stop is rolled up to the original entry position, price may move adversely to the downside and take

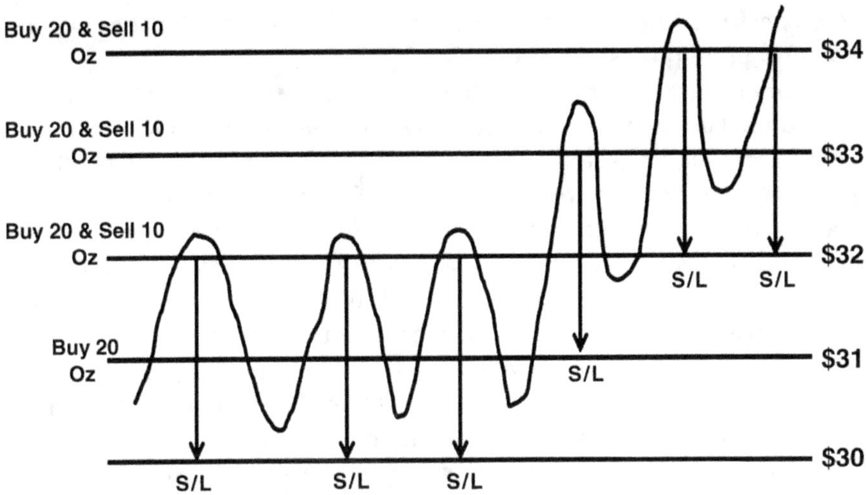

FIGURE 12.28 Taking Type 2 Profits

out the stop, thereby erasing the earlier profit exit of a Type 2 mechanism. We must avoid capital unboundedness as best we can.

In the example above we only entered new 20 ounce positions at every other level. What if we entered a new 20 ounce position at every successively higher level? See Figure 12.28. In Type 2 (scale to breakeven) with multi-interval accumulation, the vibradirectional trader enters an upbuy at every higher level at 20 ounces.

The vibradirectional trader enters with 20 ounces at $31. Just like the previous example, the trader sells 10 ounces at $32 and re-enters with 20 ounces if price retests the $31 level. The vibradirectional trader repeats every time price oscillates between the $31 and $32 levels. Accumulation at the $31 level takes place for no added capital outlay. But if the trader enters a new 20 ounces long position at $32 as the lower level position is scaled out, that trader may not be able to re-enter the $31 level with 20 ounces. If the vibradirectional trader has fully allocated all working capital at each new higher level, then no additional long positions are allowed below that level. The trader would not be able to afford holding an extra lot to ZTL. This is again the last in condition for full capital allocation per level.

But, if the vibradirectional trader decides to add more funds to accommodate a maximum exposure of 40 ounces at any one time, then the construct will allow multi-level accumulation—vibrational as well as trend capture profits!

Referring to Figure 12.33 again, we now see that accumulation and vibrational profit extraction is possible between \$31 to \$32 and \$32 to \$33, for entries at \$31 and \$32 respectively. This represents a two-level accumulation zone between \$31 and \$33. To create a three-level accumulation, vibrational, and trend capture zone, we just need to stretch our maximum exposure to 60 ounces at any one time. We start by entering a 20 ounces position at every higher level. When we reach the third level, where we buy our highest 20 ounces position yet, we will be able to accumulate at the highest two levels. But should price fall and stop out our position three levels below the highest level, the additional 20 ounces are entered. At this point, the vibradirectional trader will be holding a maximum of 60 ounces. If price stays at this level or retraces to the upside, accumulation and vibrational trading may be carried out. But if price falls four intervals, which is below the lowest 20 ounces position, then no more accumulation or vibrational trading is possible. The trader may have another downside construct to handle accumulation, vibrational, and trend capture at these lower levels.

Defensive Accumulation Technique

Defensive accumulation is an extension of the Type 2 scale to break-even exit. It may be considered a more extreme form of the Type 2 exit, whereby the vibradirectional trader scales out a larger amount in order to position the break-even stop two or more intervals below the initial scale out level.

To calculate the required lotsize to scale out in order to place an S/L (stop loss) below the initial exit level, let D1 be the distance of the required stop and D2 the distance between the entry price and the initial higher scale out level.

$$\text{Lotsize to scale out} = \text{D1}/(\text{D1} + \text{D2}) \times \text{Entry lotsize}$$

Refer to Figure 12.29.

In the example above, our original Type 2 exit scaled out 50 percent of the initial lotsize. So assuming we want our S/L to be positioned at the same distance below the entry as the initial scale out is from entry, we let D1 equals D2.

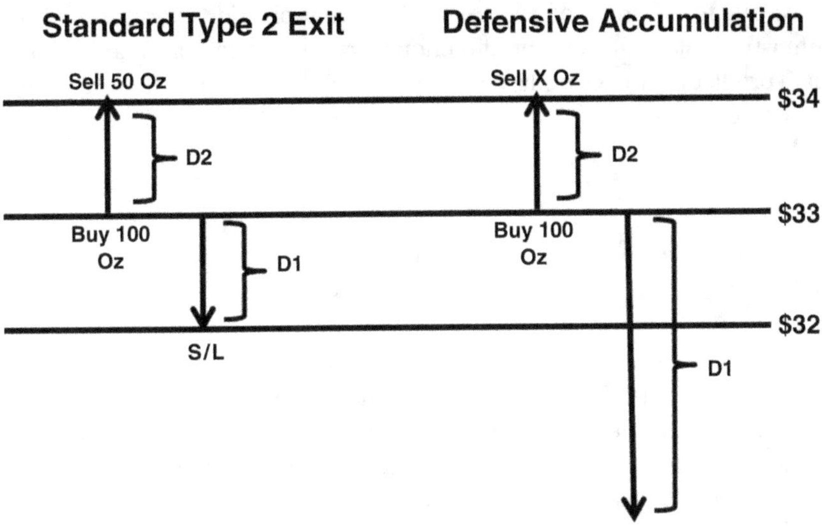

FIGURE 12.29 Defensive Accumulation

$$\text{Lotsize to scale out} = (\text{D1}/(\text{D1} + \text{D2})) \times \text{Entry lotsize}$$
$$= 100 \times (\text{D1}/2\ \text{D1}) = 50\ \text{oz.}$$

Now, if we need our S/L to be positioned 9 intervals below the initial exit, then:

$$\text{Lotsize to scale out} = (\text{D1}/(\text{D1} + \text{D2})) \times \text{Entry lotsize}$$
$$= 100 \times (9\text{D1}/10\ \text{D1}) = 90\ \text{oz.}$$

If we scale out 90 ounces across \$1, we make \$90, leaving only 10 ounces.

Should price decline nine intervals we lose 10 oz. × \$9, or \$90. Hence, the position is risk free.

The vibradirectional trader may chose to implement defensive accumulation for the following reasons:

1. To help reduce the gaps in the grid problem
2. To lock in a greater vibrational profit.

The second point should be obvious to a free swinging vibradirectional trader. As we see from the example above, the vibradirectional trader scales out for a profit of $90 for each oscillation. For every oscillation within that interval, the vibradirectional trader extracts a greater vibrational profit, only to leave a lesser amount for the subsequent capture of any trend profits.

FREE STYLING ACROSS MULTIPLE LEVELS WITHOUT RISK FREEING

Free styling is the allocation of working capital across a number of consecutively higher levels. There is no risk-freeing via Type 1 or Type 2. This is possible because the vibradirectional trader has already initiated the free styling from the capstone entry level and is therefore able to fund more constructs below the capstone for vibrational profit.

So instead of entering a 100 ounces long position at $30, the vibradirectional trader enters the minimum lotsize available at each consecutively higher level in the free styling construct.

See Figure 12.30.

Here, the free style vibradirectional trader buys the first seven levels from $30 to $38 at 10 ounces per level, allocating about $3,060 to the $8 range. To enter a higher position, capital can be freed up in the following ways:

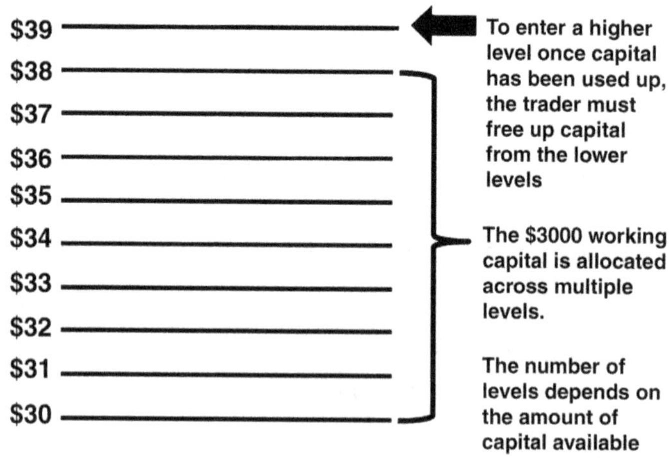

FIGURE 12.30 Free Styling across Multiple Intervals

1. Roll up each position's stop above \$31 from the lowest two levels, thereby freeing up \$300 + \$310, or \$610. This will allow the trader to enter a new position of \$39 at 10 ounces, which only requires a working capital of \$390.
2. Roll the stop from the \$30 and/or \$31 positions to a higher price level, locking in sufficient profit to cover the purchase of the new higher position.
3. Liquidate the lower levels in order to purchase the higher ones.

The Free Styling "Rolling Window" Construct

The whole idea of free styling is to capture upside trend profit while avoiding the creation of gaps in the grid. This is possible since there are no break-even stops within the free styling grid. To continue to buy into new positions, the vibradirectional trader may also execute a rolling window free styling construct where the trader does not allocate capital across the full range. Instead, the vibradirectional trader only initiates positions across three or four levels, all the while scaling out the lowest level for every new higher long entry. The vibradirectional trader will therefore have unused capital to keep this rolling window funded as price rises. The advantage is that the trader is not as fully invested as with the free swinging method, and could possibly exit the entire market with only a little loss in capital.

Free Styling with Vibrational Returns

We will now add dual sizing to the free styling technique.

In Figure 12.31 we see that if we started our first upbuy at \$31, we could exit 10 ounces at \$32 while making it risk free via a Type 2 scale to break-even exit. We would continue to accumulate risk-free positions as prices climbed, besides making vibrational profits in the process.

> *Vibrational profit does not imply accumulation but accumulation certainly does imply vibrational profits.*

If our original plan was to hold no more than 10 ounces per level, then the dual sized version only adds an extra 10 ounces to the entire construct as it progresses to the upside. The maximum lotsize held at any time is

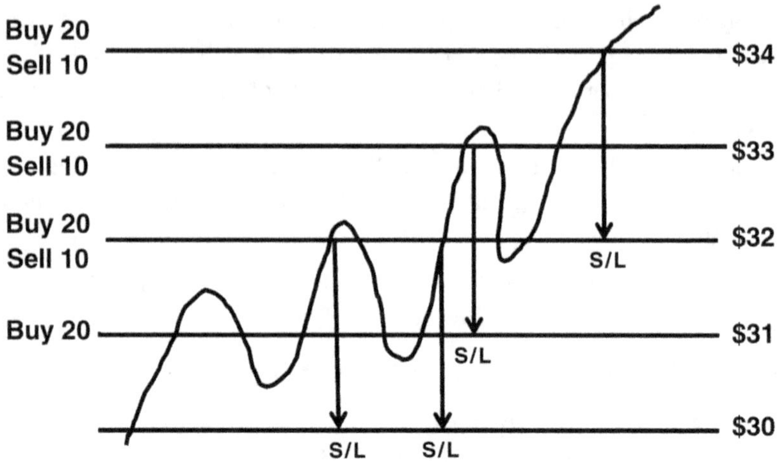

FIGURE 12.31 Free Styling with Dual Sizing

30 ounces: the remaining 10 after the upside scale out from 20 ounces, plus the new 20 ounces per level upbuy.

UNBOUNDED BIDIRECTIONAL PROFIT CAPTURE CONSTRUCTS

In this section we shall be looking at constructs that offer bidirectional profit extraction by way of vibrational returns, trend capture returns, and accumulation. The constructs will be comprised of longs and shorts. The vibratrader will now experience bidirectional vibration returns, hence the term Vibrabidirectional Trader.

Before we move into more complex constructs, we must first familiarize ourselves with the basic bidirectional mechanisms.

Bounded Short Entries

As a rule, we never initiate naked shorts within the constructs. If the market moves adversely against our position, the only solution is to exit via a stop, and in the process deplete capital. If this process is repeated enough, the trading account will eventually be obliterated.

Nevertheless, it is possible to introduce shorts into the construct if they are bounded by a long position. The entire pair must be either net long or fully hedged; no net shorts are permitted under the Principle of Capital Boundedness.

Shorts Initiated above a Long Position

Shorts initiated above a long position of equal size are totally bounded since both the longs and shorts are entered via limit orders. This is in line with order entry boundedness. Should price fall, the short may be covered for a short profit, and thereafter re-shorted. But if price rises, the vibrabidirectional trader will still lock in a profit from the pair. To exit a hedged pair, the vibrabidirectional trader must exit both the longs and shorts simultaneously. The longs must liquidate with a position or lotsize greater or equal to the shorts.

According to the principles of price action, as seen in an earlier chapter, the subsequent direction of price is irrelevant to the pair as the profit has been soft-locked. Soft-locking means that some profit or loss has been captured by two opposite open positions. Though there is profit in this case, it has not been monetized or realized as yet, and is still in the form of a hedged open position in the market.

Let us say that two positions were initiated, one long position of 100 ounces of silver and another with a 70 ounces short position. Of course if price declines, the short can be covered for a profit since the long was part of the original construct and the shorts only represented a bonus should the best-case scenario unfold. But if price rose and only the shorts were exited, it would cause capital erosion should price fall afterwards. If the process is repeated enough times, all the trading capital would be eroded away permanently.

Shorts Initiated below a Long Position

Generally if a short is initiated below a long position of equal size, it is considered an unbounded situation. If price falls there may be short profits to be made. If it rises, the losses have been soft locked into the hedged pair. The only way out is to only initiate a short below a long position (if there is another long above the short or if the original pair was net long). Then all we need to do is wait for price to rise enough to cover the soft-locked losses before exiting or transforming it into another mechanism.

THE "BIG HEDGE" TECHNIQUE

The big hedge is a long-short construct whereby the vibrabidirectional trader extracts vibrational profit from both long and short price oscillations. It is the purest bidirectional mechanism.

Figure 12.32 displays a long-short construct. This construct has only one purpose—to extract as efficiently as possible all vibrations between $30 and $20. It is not designed for any upside profits, though you may always make the top end net long with fewer shorts.

There are two conditions that the vibrabidirectional trader must accept:

1. That the total long positions held below $20 must not exceed 200 ounces
2. That this construct is not designed for upside trend capture.

The 100 ounces long position is placed at that level so as to keep the upside hedged against the 100 ounces of shorts. Remember, the vibrabidirectional trader will have planned another construct below and above this construct. The 10 ounces seen on the left-hand side illustrates the regular long downside construct that was already planned. The 10 ounces shorts are exited in one interval move starting from the first one on the left. The first short position exits at $29. The second exits at $28, and so on. But

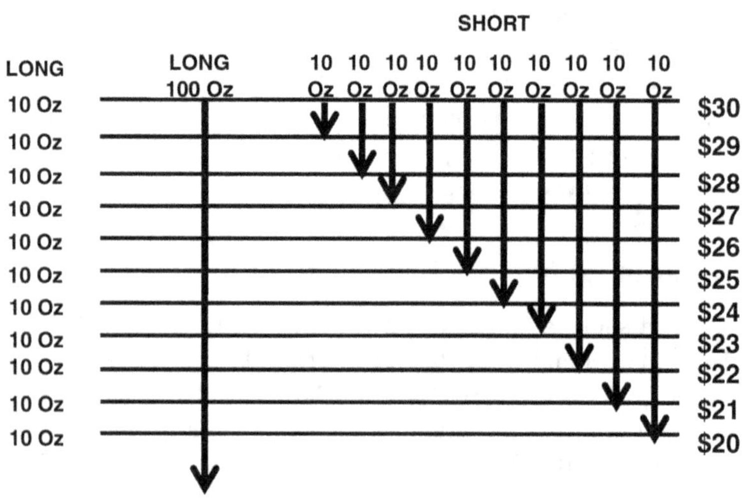

FIGURE 12.32 The Big Hedge

each short exit must be re-entered once price retraces up by one interval. This means that the first short, which exited at $29, must re-enter if price rises back to $30. The second short position must re-enter its short at $29, and so on.

So, when price falls from $30 to $29, the short exited with a profit. When it rebounded back to $30, the short was re-entered, ready to repeat it all. We notice that the vibrabidirectional trader earns in the downside as well as the upside. This long-short profit extraction is only available within the $30 to $20 range that we planned. This bidirectional profit extraction boosts the returns by 100 percent. If you find that your instrument of choice has a certain average daily range and your intervals are designed around this range, you will see much better results with this construct.

Normally in standard unbounded trading the term bidirectional means placing a buy above and a short below, encapsulating price and waiting for a profitable breakout in either direction. It still exhibits oscillational losses should price whip across the long-short zone. But our construct represents pure bidirectional trading.

THE "SMALL HEDGE"

The small hedge is illustrated in Figure 12.33.

We start off by being long 20 ounces and short 10 ounces at $30. At $29, we cover the short for profit but enter a new long for 20 ounces. If price rises, we re-enter the short at $30 and exit the long with profit by

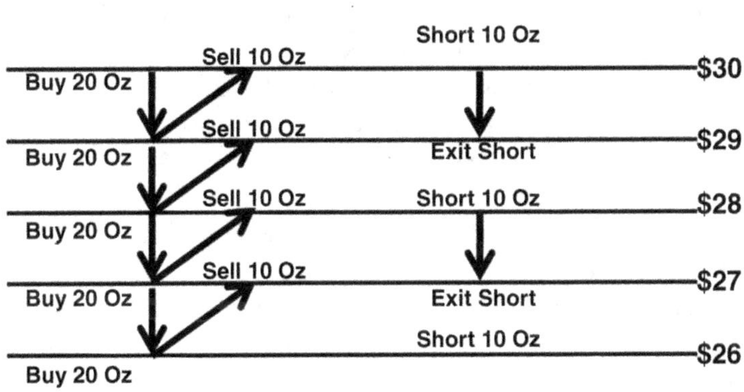

FIGURE 12.33 The Small Hedge

liquidating 10 ounces from the \$29 level. So, as you can see, this \$20 to \$30 level extracts price vibrations in two directions; it is a pure hedge. We do not have short positions anymore at \$29, and we proceed to short 10 ounces at \$28 and long another 20 ounces at the same price. If price rises back to \$29, we may then liquidate the long position from \$28. The shorts are still hedged by the longs at \$20 and \$30. Hence, at this level, the oscillational returns are based solely on the longs. If price falls to \$27, the whole sequence from the \$30 to \$29 interval repeats itself at the \$27 to \$28 intervals.

The small hedge is very light and easy to implement. It provides pure vibrational extraction at every second interval and standard vibrations in between. Many have found the small hedge to be particularly useful in conjunction with the non-dual sized, non-accumulation-based construct described earlier, which is just a simple upbuy-upsell-downbuy-upsell construct. These two are perfectly interchangeable and work seamlessly to track the market to the upside and downside with a minimal of capital.

THE UPSIDE SHORT HEDGE

So far, the vibrabidirectional trader is only able to extract short vibrational profit to the downside. In order to extract short upside vibrational profits we must first make sure that the construct is not exposed to unlimited losses when shorting to the upside. We can easily accomplish this by creating an artificial market top to cap the short losses.

First we find out the number of upside short levels we intend to initiate. We then calculate the cumulative intervals. We must also initiate a larger long position at the entry level so that at the highest short level, the long position profit cancels out the cumulative short losses.

Let the number of upside short levels be 10, the distance between the short intervals be y.

Let n be the size of the short positions (assume equal sizing per level) and m be the single larger long position size.

The number of cumulative intervals will be $(10 \times 11)/2 = 55$ intervals.

$$\text{Long profits} = m \times \text{Distance to highest short level} = (m \times 10y) = 10my$$
$$\text{Cumulative short losses} = (n \times \text{Cumulative intervals}) \times y = 55ny$$

To balance out, the long profit and short losses must match up at one interval above the highest short level, therefore:

$$\begin{aligned} \text{Long profit} &= \text{Cumulative short losses} \\ 10my &= 55ny \\ m &= 5.5n. \end{aligned}$$

The solution, $m = 5.5n$, is called the long/short position ratio. This ratio varies for different numbers of intervals, but it does not vary with the interval size. Notice that y cancels out and therefore this ratio is independent of interval width. Assuming that our shorts are 10 ounces each, the long position required to cap the losses at one interval above the highest short position is 55 ounces, which we call the termination level. This level represents the break-even point of the entire construct. This construct allows the vibrabidirectional trader to extract upside vibrational profit without the prospect of unlimited losses. But the trader must exit all positions, long and short, at the termination point. Otherwise the construct becomes unbounded and is exposed to upside losses. Making the long position a little larger will usually mitigate any small losses at the termination exit due to gapping or slippage. The construct is actually net short above the short positions average price. One significant advantage of the upside short hedge is that once price has risen above the shorts average price, any moves below the original entry level will yield short trend profits if the shorts are held. The long and short can both exit at any price below the entry level.

ZERO COST HEDGING TECHNIQUE FOR "LOADING THE MATRIX"

There is normally no advantage derived from a "perfect hedge." But once we trade in a bounded manner, the perfect hedge may be of utmost importance. See Figure 12.34 for an illustration of a Zero Cost Hedge construct.

In bounded trading we load up the matrix first with these zero cost hedges. Sometimes the vibrabidirectional trader is hesitant to invest in a particular market and loads up the ETF or CFD until it looks advantageous to participate. Should price fall to the lower levels in an instrument that cannot test ZTL, we can take advantage of the low prices to make an entry by releasing all shorts. In effect, the shorts have paid for our running capital (or cost) and left us with a convenient upside vibrational grid for further profit extraction.

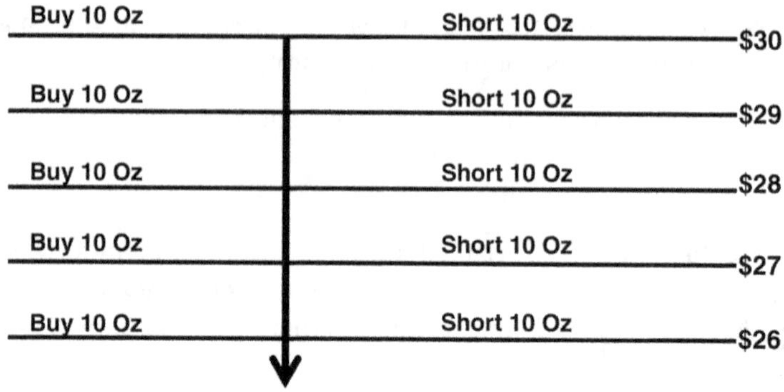

FIGURE 12.34 Zero Cost Hedging

But before releasing all or some of the shorts, we need to make sure that we have enough margin and capital to hold all the unhedged longs we have created. We can also implement zero cost hedging during market declines and later reap the benefits of a reduced running capital, which you most likely would have funded had you not used the hedge. You have the opportunity to get in on hindsight without making a commitment to the market. The disadvantage of the zero cost hedge is that it is of little use to the upside in bounded trading. It also prevents the vibrabidirectional trader from earning any vibrational returns while the hedge is in place.

Vibrahedging Technique II

Vibrahedging 1 was covered in an earlier chapter. But in this version, we initiate Macrosiso Vibrahedging with a zero cost hedge construct. As price falls to the next interval, we cover the shorts and fund a simple bi-directional breakout mechanism, just as we did in the first vibrahedging method. The only difference here is that the bidirectionals are funded by the short and not the long vibrational profits.

MORE CONSTRUCTS

Linear Downbuy-Upsell Constructs with a Scale Factor of 1

The most basic construct a vibratrader can build is a simple linear down-buy-upsell matrix or grid where all positions and intervals are of equal size.

Any construct where all the shares or lotsizes are of equal size has a scale factor of 1. These are the Microsiso/Macrosiso type constructs.

The simplest vibrational and trend capture mechanism to the upside is a non-dual sized upbuy-upsell-downbuy-upsell mechanism. You enter a 100 share position at $56, for example. You proceed to liquidate the shares for a profit at one interval above the current price. You then re-enter the 100 share position when price retests the $56 level, followed by another profitable exit at the next level above—basically a standard vibrational trade. The only issue is that your next upbuy will only be available after the upsell.

Martingale Downbuy-Upsell Constructs with a Scale Factor > 1

Martingales are one of the most powerful constructs within a bounded trading environment. It is definitely not recommended for use in unbounded trading systems.

The best type of Martingale will be one that clears all running capital in a single upsell. As you can see from Figure 12.35, the upper running capital was clear at points 1, 2, 3, and 4. And with a ZTL that is virtually impossible to hit, we have the perfect set-up for vibrational profit extraction. In fact, we hope that prices decline so that we may access the lower levels of our

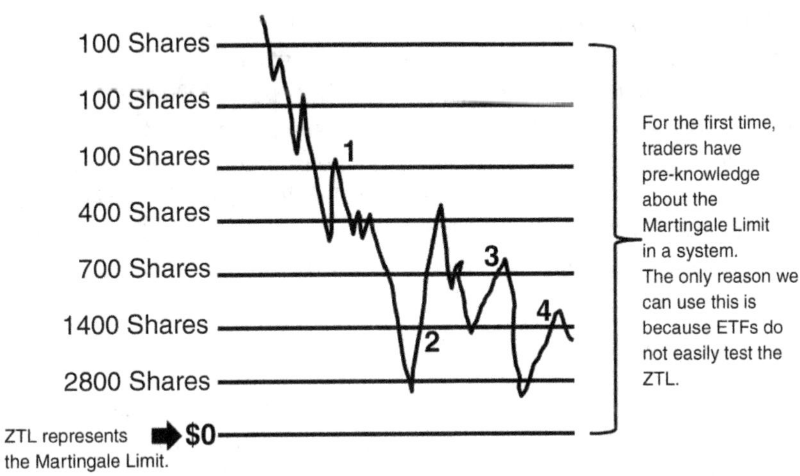

FIGURE 12.35 Martingale Based Constructs

Martingale vibrational construct, where the returns are greatest. The only reason all this is possible is because a ZTL event is highly improbable.

Remember that the big boys of trading require very little skill to extract profits from the markets because they have an inexhaustible supply of funds. As a result, they can buy as prices are falling and sell as they rise, whereas most traders exit the market in a decline and enter the market when it rises. Vibratrading puts you in the same position as these big boys for the very first time!

Down-Shifting Technique

Should price retrace to the downside and access the lower levels of your construct, you may also opt to exit the entire construct and re-initiate at a lower level. This is called down-shifting your positions. The only disadvantage is that you will lose your downside Microsiso and Macrosiso vibrational constructs. Therefore if prise rises, you will need to use either the free swing or free style techniques to gain exposure to the upside.

Single Upsell Technique for Managing Downside Running Capital

The single upsell technique, which has been covered in an earlier chapter, also represents an exit strategy from the entire traded instrument.

Let us say that you experienced $200 running capital (or cost) in a downside vibrational construct. To clear this downside running capital, you may buy 200 ounces (or shares if it is an ETF) to gain a $200 return from the interval oscillation. If that is too heavy a position to enter, you may use a smaller lot or sharesize to gradually clear the running capital. Each oscillation with 50 ounces or share positions will clear 25 percent of the cost. Even if price should fall below the single upsell mechanism, you will have already accounted for that scenario and planned further action at lower levels.

Horizontal Reinvestment

Should the traded ETF ever be delisted, or should a CFD broker wind up, we may get the opportunity to horizontally reinvest the working capital in a new construct with a lower PLR if possible. This will help the vibratrader restart the vibrational constructs.

The Visual ATR (1) Method

Before setting the interval sizing for the vibrational constructs, we need to at least be familiar with the average period range of the traded instrument in question. One way you can do this is through the single period Average True Range (ATR), or ATR (1). Apply the ATR (1) to various time frames and place a horizontal line across it above about 90 percent of the ATR (1) values. Make a note of the ATR (1) value at the level of your horizontal line.

Let us say you discovered that silver has an average daily range of $1. You would then design and plan your intervals around this size so that you can, on average, expect at least one interval profit per day. This is more probable for pure bidirectional constructs like the big and small hedge because a $1 move may be up or down. If the construct extracts long as well as short vibrations, then it would tend to produce on average one hit a day.

The Implications of UPLR in the Vibrational Matrix

What is the percentage profit for a $1 rise in price from $30?

If we long 10 oz. at $30, then our percentage return over working capital will be:

$$\begin{aligned}\text{Percentage return} &= \text{Profit/Working capital} \times 100\% \\ &= (10/300) \times 100\% = 3.33\%.\end{aligned}$$

Increasing the lotsize to 100 ounces yields:

$$\begin{aligned}\text{Percentage return} &= \text{Profit/Working capital} \times 100\% \\ &= (100/3{,}000) \times 100\% = 3.33\%.\end{aligned}$$

The results are identical. In fact, there is no way anyone can earn more or less than 3.33 percent at $30 for a $1 interval move.

The percentage profit is actually the Unit Price Leverage Ratio or UPLR, and is defined as UPLR = 1/Entry Price.

So the highest and lowest percentage you can make per $1 rise from $100 is 1/100 or 1 percent.

Since we have no control over the percentage profit except to select an instrument that is very low priced, we may still base our calculations on the absolute returns. For example, if you wanted to make \$100 per \$1 move, all you need do is position 100 ounces at any price level and allow it to rise \$1.

Designing Vibrational Constructs to Extract a Fixed Percentage per Average Period Range

We use the ATR to determine the interval size we require in order to extract the desired profit over a specific time period. Let us say that you want to extract daily profits of around 1 percent.

Let us also assume that the daily ATR for silver is \$1. Assume that we design our construct to buy from \$30 to \$29, using a \$1 interval. Let x and y be the ounce per \$30 and \$29 level respectively.

To extract 1 percent per oscillation per level:

$$\begin{aligned}\text{Total working capital} &= (100\%/1\%) \times \text{Profit per interval } 30x + 29y \\ &= (100/1\%) \times \$1\end{aligned}$$

Therefore $30x = 71y$.

Assuming our initial lotsize allocation, that is, $x = 10$ oz.:

$$y = (1/71)(30x) = 0.422 \times 10 = 4.22 \text{ ounces.}$$

Therefore, to earn 1 percent we need to place a lotsize of 4.22 ounces at \$29.

Proceeding lower to the \$28 level (let y be the lotsize at \$28 and using $y = 30x/71$ from above):

Total working capital for all three levels $= 100y$

$$30x + 29/71(30x) + 28y = 100\,y.$$

$$y = (1/72)\,(30x + 29/71(30x)) = 5.86 \text{ oz.}$$

Therefore, to earn 1 percent per oscillation from \$28 to \$29, if our lotsize at \$30 is 10 ounces, our new lotsize at \$28 will need to be 5.86 ounces.

At the \$27 level: (let y be the lotsize at \$27 and using $y = (1/72)\,(30x + 29/71\,(30x))$ from above):

$$30x + (29/71)\,(30x) + (28/72)\,(30x + 29/71\,(30x)) + 27y = 100y$$

$$y = (1/73)\,(30x + 29/71\,(30x) + (28/72)\,(30x + 29/71\,(30x)) = 8.02 \text{ oz.}$$

To earn 1 percent per oscillation from \$28 to \$29, our new lotsize at \$28 will need to be 8.02 ounces.

As we can see, there is a repeating pattern here.

- \$30 level = 10 oz.
- \$29 level = 300/71 = 4.22 oz.
- \$28 level = 422/72 = 5.86 oz.
- \$27 level = 586/73 = 8.02 oz.
- \$26 level = 802/74 = 10.84 oz.
- \$25 level = 1084/75 = 14.4 oz.

$$\text{Initiating formula} = \text{Capstone capital}/((100/\text{Percent return}) - \text{Price level}).$$

Hence to earn 1 percent per day, according to the daily ATR of \$1 per interval, we need to allocate a specific number of ounces as we proceed to lower levels. Note that 1 percent extraction per \$1 intervals is not possible for prices above \$100 as the UPLR will be too low.

We may also implement the big hedge method in order to extract at 100 percent efficiency.

EXITING WITH PROFIT

Exiting With Vibrational Returns

The vibratrader exits with either downside or upside vibrational (or scalping) returns.

Exits are usually at the next upsell for Microsiso but are at two or more intervals for the Macrosiso mechanism. A trailing stop may be used and positions are left to run unhindered.

The Macrosiso positions can be exited as per the trend-based returns, as seen below. The Last In First Out, or LIFO Exit refers to exiting a position profitably during a reversal in price. The position is exited one interval before being hit by price action. Normally, during an uptrend line break, positions are exited for a one interval profit, as price falls from the trendline break-out to the downside.

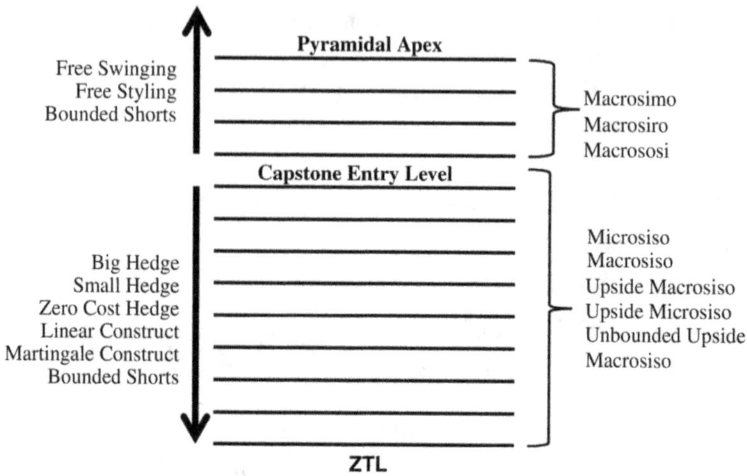

FIGURE 12.36 Summary of Constructs

We may exit in the following ways:

1. Initiate a trailing stop at four to five levels above initial entry level, positioned at a specific number of intervals behind current price.
2. Exit a few lower-level entries arbitrarily to lock in some profit.
3. Exit 50 percent of positions when a major trend line is breached and initiate LIFO Exit Mechanism (dual sizing required).
4. Exit 50 percent at next upsell level (dual sizing required).

See Figure 12.36 for an illustrated summary of the concepts covered in this chapter.

CHAPTER 13

The Vibrational Vehicles

Vibratrading employs equity ETFs and commodity-based CFDs as its main trading and investing vehicle. ETFs offer an exceptional way to access various local and global markets, and they put traders and investors on equal footing with larger hedge funds in terms of being able to participate in all of the following markets with minimal capital:

1. Industries
2. Sectors
3. Futures/commodities
4. Local/international stock indexes
5. Global markets
6. Emerging markets
7. Specialty markets
8. Style (growth, value, small cap, large cap)
9. Bonds/fixed income
10. Currency.

Contract For Difference (CFDs)

CFDs provide the vibratrader with many unique advantages that equity ETFs do not offer. High leverage and access to twenty-four-hour trading of spot-based commodities are but a few advantages. Nevertheless, we only trade commodities when using CFDs, as we want to reduce the incidence of a zero test level event. There are many splendid books out there on CFDs and ETFs, and this book does not purport to deal with the intricacies of a subject that would take an entire book of its own to cover. Only a brief description of ETFs and CFDs will be given.

CHARACTERISTICS OF EXCHANGE TRADED FUNDS (ETFs)

In the majority of ETFs, you are able to:

1. Buy a minimum of one share
2. Buy on margin (just like regular stock)
3. Short the ETF (if shares are available)
4. Buy and liquidate intraday (subject to the Pattern Day Trader's Rule)
5. Trade ETFs with real-time prices and not end-of-day prices
6. Use all traditional order types (limit and stop order, etc.)
7. Receive dividends (shares are held by the brokerage on behalf of client).

Since ETFs represent a basket of stocks, they are valued via their Net Asset Value, or NAV. The NAV is kept fairly close to the actual value of all holdings by virtue of arbitraging.

The ETFs Built In Protection Mechanism

The most important feature of an equity ETF is its stock replacement mechanism, which replaces any stocks that fail to satisfy the ETF's stipulated standards for inclusion into its basket of component stocks. Hence, the value of the shares of an ETF will very likely never fall to zero.

It is for this reason that vibratrading chooses equity ETFs as its main vehicle for bounded global market exposure.

ETF Categories

There are over 800 ETFs currently in the market. Here is a list of the main ETF categories available to traders and investors:

- Broad market
- Large, mid, and small cap
- Growth and value
- Currency
- Commodity
- International
- Country
- Developed market
- Emerging market.

ETFs, ETNs, and Holders

Exchange Traded Notes (ETNs) are in practice more like debt instruments where the issuer promises to pay the holder any profits accrued. Should the issuer go bankrupt, the holder may not be able to realize any value in the notes. The main difference between ETFs and ETNs lies in the nature of their obligation to the holders. Therefore, the vibratrading methodology only focuses on equity ETFs (rather than ETNs) to reduce exposure to issuer default risk.

TYPES OF RISK ASSOCIATED WITH ETFs

In order to maintain the foundations of a strong construct, we need to ensure that our trading vehicle has the least susceptibility to:

- Liquidity risk
- Systematic risk

- Asymmetric leverage risk
- Specific risk
- Interest rate risk (for bond ETFs)
- Default/counter party risk
- Execution risk.

Liquidity Risk

We only trade stocks that are of sufficient daily volume. Many stocks do not have the minimum volume of trading to ensure efficient fills and tight spreads. But the volume creation process associated with an ETF implies that the trader will be able to trade many ETFs in spite of lower volumes. The liquidity of an ETF depends on the collective liquidity of all its component stocks. Nevertheless, many popular ETFs trade above 100 million shares daily.

A quick check via various online finance resources will sort out the most active ETFs by volume.

Specific Risk

There is always a danger that a stock may cease trading or sustain unprecedented declines in market value due to some internal problem specific to the company, for example, missed earnings, scandals, fraud, or mismanagement. Hence it is very risky to invest a lot of capital into a single stock. An ETF is a basket of stocks and therefore has a diversification advantage. Total decline in any one or two stocks will not impact the overall NAV of the ETF too drastically, even if that stock is a large percentage of its holdings. ETFs do replace component stocks should any fail to meet its criteria. This feature, in combination with the inherent diversification, will keep the NAV afloat.

Default/Counterparty Risk

Just as stocks are subject to specific risk, ETFs are sometimes subject to termination and liquidation (de-listing from the exchange). This is rare, and generally occurs when the issuer deems that there is insufficient trading or participation in the ETF. When an ETF liquidates, the ETF stock holders will receive cash payment equivalent to the final NAV upon liquidation.

They may also opt to receive its equivalent value in some of the ETF's component stock.

There is also some fear as to the integrity and financial stability of brokerages, especially when FOREX and CFDs are involved. Vibratraders must take into account the financial standing of various brokerages and the level of their risk capacity with regard to counterparty risk.

FUNDS TO AVOID IN VIBRATIONAL TRADING

You should never use the following funds in vibratrading:

- Merrill Lynch HOLDRS: These are fixed baskets of stocks created for trading of "speculative asset classes" rather than for balanced and diversified portfolio exposure. HOLDRS become less representative of their asset class over time and hence more volatile.
- ETNs
- Leveraged ETFs.

Vibratrading depends on the trading vehicle preserving total allocated capital in the face of severe market or systematic risk. The pyramidal foundations will not hold up in the unlikely event that the NAV of an ETF declines to zero. You may replicate an ETF by buying multiple single company stocks, however, this is the only time buying single company stocks is allowed.

Leveraged and Short ETF and ETN Funds

Leveraged ETFs provide returns in excess of 100 percent. They utilize a range of underlying financial products to achieve this extraordinary return, ranging from swaps and options to futures contracts. Two very popular leveraged funds are the FAS and FAZ, which return 300 percent with respect to RIFIN (Russell Financial Services Index), their underlying index.

Short funds (leveraged and unleveraged) and long leveraged funds are similar in their underlying tracking behavior. They display asymmetric leverage, meaning there is a short bias "built in" with returns suffering negative expectancy. This negative expectancy is the direct result of the daily reset mechanism and its interaction with price oscillations.

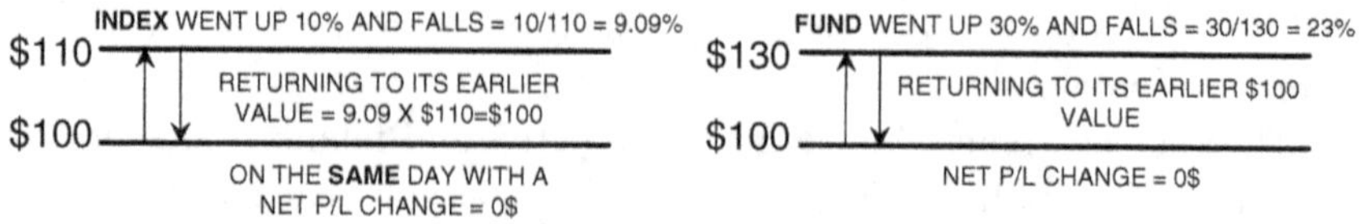

FIGURE 13.1 No Daily Reset Feature for Intraday Moves

Leveraged and short funds are reset daily, so the returns are based on the difference between the current daily price and yesterday's close price. It does not track the underlying index on a continuous basis. Let us examine the implications of the daily reset feature to further explain why we only trade unleveraged ETFs.

Assume price of the underlying rises 10 percent with respect to yesterday's close. The leveraged ETF rises 30 percent (since it gives a 300 percent return based on the underlying index). On the same day, the index falls 9 percent back to $100. The fund falls 23 percent back to its original price of $100. The fund has tracked to underlying throughout the trading day and has replicated its moves by a magnified level of 300 percent, as seen in Figure 13.1.

Price of the underlying rises 10 percent from yesterday's close. The leveraged ETF rises 30 percent. The next day the index returns to its earlier close, falling 9 percent back to $100. The leveraged fund falls 27 percent (9 percent × 3) back below its original price of $100. Refer to Figure 13.2.

As you can see, even when the index retraced to starting price, the fund experienced negative expectancy and fell below its starting price.

Now consider if the price of the underlying falls 20 percent with respect to yesterday's close. The leveraged ETF falls 60 percent. The next day the index returns to its earlier close, rising 25 percent to $100. The leveraged fund rises three multiplied by 25 percent, or 75 percent, but it still below its original price of $100 (see Figure 13.3).

As you can see, even when the index retraced back to starting price, the leveraged fund experienced negative expectancy during price oscillations and fell below its starting price by over 65 percent.

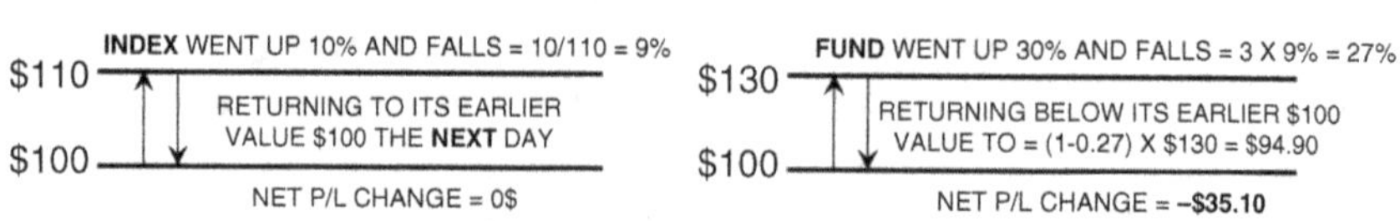

FIGURE 13.2 Daily Reset Affecting Interday Moves

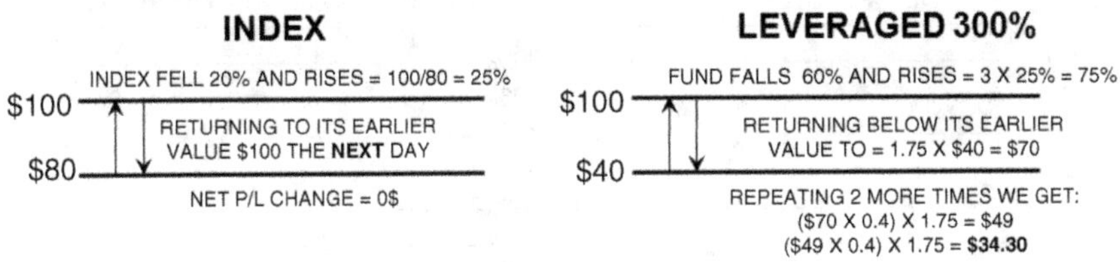

FIGURE 13.3 Negative Expectancy after Rising to Initial Price

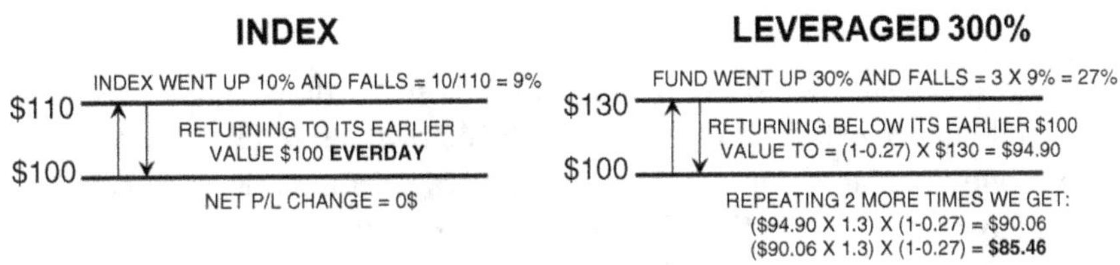

FIGURE 13.4 Negative Expectancy after Declining to Initial Price

Now let us repeat this three times at a change of 10 percent, each time having the underlying fall back to its starting price of $100.

As you can see from Figure 13.4, even when the index retraced to starting price every day for three days, the fund experienced tremendous negative expectancy and lower price with every oscillation.

Figure 13.5 gives an example of negative expectancy in a short ETF displaying the same characteristics.

When the index retraced to starting price, the short fund experienced negative expectancy during price oscillations and fell below its starting price, even with a leverage factor of one.

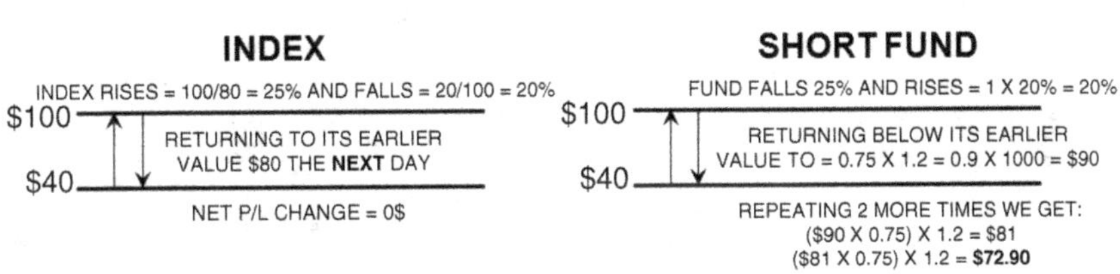

FIGURE 13.5 Negative Expectancy in Short Funds

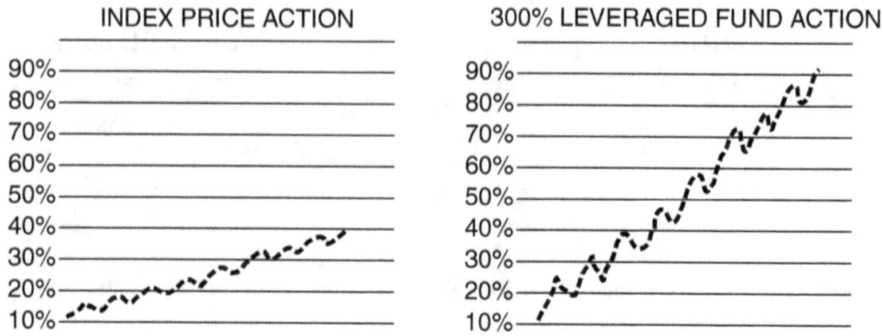

FIGURE 13.6 Leverage Fund Works Best in Trending Markets

As you can see, leveraged and short funds suffer drastic price declines during oscillations. It is similar to issues faced by fixed fractional and fixed ratio money management systems. But leveraged funds do respond phenomenally to trending action. A three day rise of 10 percent per day in the underlying will return about 120 percent. Similarly, a fall of 10 percent per day for three days will lead to about a 66 percent fall in price from a 300 percent leveraged fund (see Figures 13.6 and 13.7).

Tracking Error

Another area of concern in dealing with ETFs is the level of tracking accuracy between the ETF and its underlying instrument. Generally, equity ETFs will track their underlying basket of stocks very closely, as the SPY does in Figure 13.8.

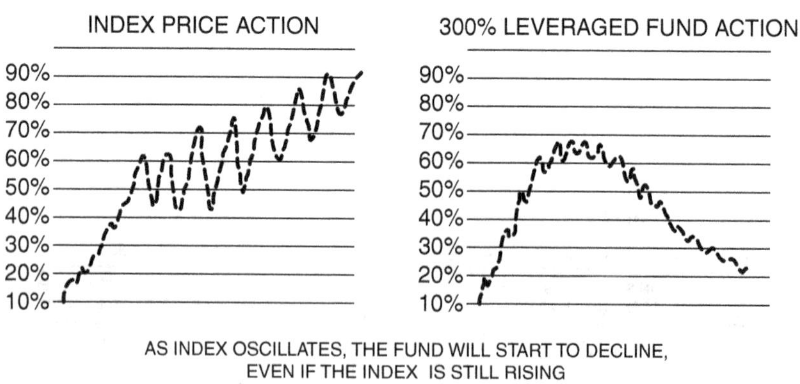

FIGURE 13.7 Leverage Fund Underperformance in Ranging Markets

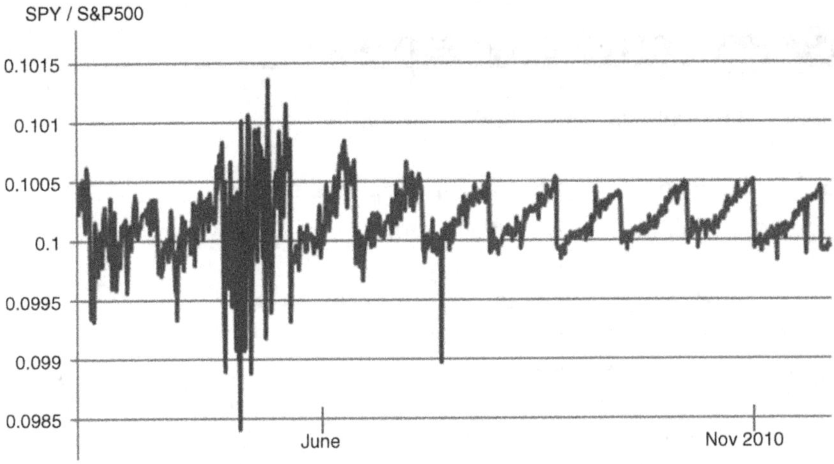

FIGURE 13.8 Performance of SPY to its Underlying Index

Unfortunately there are many ETFs that cannot track their underlying accurately. In many cases the underlying futures contract has a negative roll yield, probably resulting from contango in the market, which produces losses in the ETF tracking it. This is usually seen in commodity-based ETFs and ETNs. Figure 13.9 illustrates this underperformance clearly in UNG, which tracks natural gas.

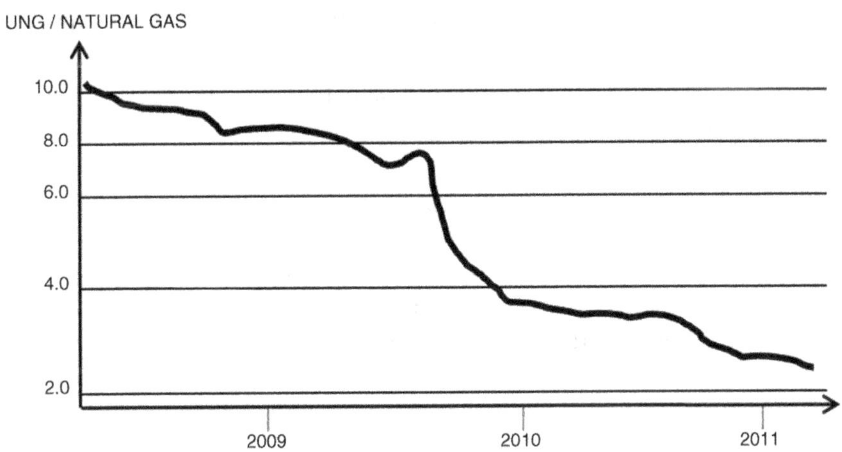

FIGURE 13.9 Underperformance of UNG to Natural Gas

THE REPLICATED ETF PORTFOLIO

Vibratrading's utilization of ETFs is primarily for:

1. Global exposure to otherwise inaccessible markets
2. Risk reduction via diversification of invested equity exposure.

But if global exposure is not a priority in your portfolio, then a Replicated ETF Portfolio (REP) is a viable option.

Vibrational traders may use the oscillatory propagation technique should they intend to trade single stocks. They could use this technique to create free stocks from market oscillations within their ETF replicate portfolio. By holding a large and sufficiently diversified equity portfolio, they replicate what the ETF primarily provides, except that they will not experience any price averaging.

Replication has three major advantages over ETFs as far as profit extraction is concerned:

1. Holding individual stocks exposes your portfolio to greater volatility, which is ideal for bounded vibratrading. (ETFs are naturally less volatile than stocks due to heavy diversification and the way share price is averaged, which can dampen vibrations.)
2. Optimization of the scaling mechanisms can be performed and fine-tuned on each stock individually for increased volatility matching and profitability.
3. It is easier to locate higher PLR and BLR stocks as there are many more stocks priced lower than ETFs.

Also, ETFs do have some disadvantages, namely:

1. Expense Ratios. Trading and investing individual stocks incurs no extra cost above trading costs. But ETFs do charge a minimal fee for maintaining and managing the fund, called the expense ratio. The annual fee is generally very low, in the region of 0.5 percent of the total investment.
2. Cost of entry and BLR. ETFs are generally more expensive than stocks, so less capitalized traders may find it more practical to

concentrate on primary market exposure. With cheaper stocks, BLR and potential returns increase.

A compromise would be to invest in a combination of ETFs and single stocks, the percentage of exposure being determined by your unique risk capacity. Vibrational profits from stocks, which give a larger and more direct exposure to the source of the vibrations, are usually greater than vibrational profits derived from an averaged or damped basket of stocks like an ETF.

CHAPTER 14

Comparison with Other Trading Systems

VIBRATRADING VS. SCALE TRADING

Conventional scale trading is in fact the deployment of the Microsiso mechanism itself, whereas averaging down is more akin to the Macrosiso mechanism. As you know by now, vibratrading is the effective use of both scale trading and averaging, offering greater flexibility and performance. Various simple mechanisms like dual or multiple sizing help the vibratrader extract both trend and vibrational returns simultaneously.

VIBRATRADING VS. DOLLAR COST AVERAGING

Dollar cost averaging is one of the most popular trading and investing methodologies around, and is heavily used by many mutual funds. Its drawback is that it continues to invest in a market regardless of price. Even though it does purchase fewer shares as price rises, it will start to focus too much funding in the market at a particular price level should the market experience a very long period of consolidation. Vibratrading avoids this pitfall by only buying and selling when there is price motion in the markets. Finally, dollar cost averaging does not really have an exit strategy, and could thus prove disadvantageous should price revert to much lower levels.

VIBRATRADING VS. VALUE AVERAGING

Value averaging is slightly more responsive than dollar cost averaging. It has the advantage of occasionally taking some profits off the table, which makes it a much safer methodology, especially in uncertain financial climates. Nevertheless, value averaging does not have vibratrading's advantage of being able to profit from even flatlining markets, not to mention the flexibility to extract short-term vibrational profits while waiting for longer-term capital appreciation in the form of trend profits.

VIBRATRADING VS. BUY AND HOLD

Buy and hold traders really miss out on profit-making opportunities by just waiting, holding, and hoping that their investments will eventually rise above the initial invested value. While the investment is experiencing drawdown, there is no form of income available except maybe some dividends. Many buy and hold investors eventually resort to averaging down to lower the average price of their stock. They fail to reduce the cost basis of their investment, which can only be done effectively via vibrational extraction.

VIBRATRADING VS. DIRECTIONAL TRADING

The true power of vibratrading lies in its ability to extract trend as well as oscillation profits from market action. Directional traders must constantly worry about their expectancy and reward to risk ratios. Every losing trade depletes capital. But in vibratrading, directional profits may be extracted without ever experiencing any capital depletion or bleed. This is the greatest advantage bounded vibratrading has over pure directional trading. In fact, one can even trade economic announcements from the safety of the capstone base, benefiting from all the price gapping and volatility.

CHAPTER 15

Profiting from Non-Vibrational Flatline Price Action

Vibratraders can usually extract oscillational profits from ranging or sideways markets. They may experience some difficulty should the interval size be larger than the current price oscillations in the market. This can be somewhat mitigated by introducing short options into the construct, thereby generating returns from the premiums instead.

THE BASIS FOR NON-DIRECTIONALITY

With the collection of premiums within the construct, the vibrational trader can look forward to a more consistent income stream. Price need not even oscillate to derive a profit from within the construct.

One disadvantage of equity options is that they are priced in a lot of one hundred shares. Therefore, the trader must realize that the minimum downbuy sharesize will be one hundred shares, which limits the minimum capital required to build a vibrational construct. Another limitation is the inability of the vibrational trader to match up strike prices with various upsells and downbuys as strike prices are only available at specific price levels.

RISKLESS SHORT OPTIONS TRADES

Once they are introduced into the vibrational construct, all short options become essentially riskless. This is due to two factors:

1. The short calls are covered by shares purchased at previous lower downbuys. If price rises above the option's strike price, the vibrational trader will still derive a profit, keeping the short call premium while earning the difference between the downbuy and the strike price.
2. The naked short puts will generate additional profits as long as the puts are not assigned. Even if the short puts are assigned, the vibrational trader would have essentially been paid to initiate the downbuys. It is a win–win situation.

USING SHORT OPTIONS IN VIBRATRADING

Short options are very easily implemented within a vibrational construct. First, sell a put at the desired downbuy level. If it is not assigned, repeat the process and collect premium indefinitely. If it is assigned then simply buy the shares at the downbuy level and keep the premium. Then, sell a new put at the next lower downbuy level and sell a call at the first upsell level. You are at this point fully covered by the long position. You may also try selling a really far OTM put at successively lower downbuy levels and collecting any premium available, no matter how small. Once initiated, you can repeat this process indefinitely. The vibratrader is essentially being paid to initiate downbuys and upsells!

As seen above, if price fails to penetrate short put strike levels, the vibrational trader need not even initiate a downbuy in order to generate a profit. But, the call must never be sold until a downbuy has been triggered.

CHAPTER 16

Summary of Vibratrading

Markets move up, markets move down. That is their nature. No matter which underlying forces drive these fluctuations, traders and investors will always be confronted with the challenge of balancing reward and risk. But the vibratrader need only be concerned with the challenge of balancing reward and opportunity.

THE TWO RULES OF VIBRATRADING

In a nutshell, vibratrading only works if:

1. There is virtually no possibility that the traded instrument or market can test ZTL.
2. There is sufficient, disposable, and timely capital for the implementation of the vibrational constructs.

The first point is of immense importance. We only trade equity ETFs and commodity-based CFDs. Single stocks are avoided as there is no mechanism preventing a single stock plummeting to zero test level and winding up (unless we replicate an ETF or use the oscillatory propagation technique).

If the two basic conditions above are fulfilled, the vibratrader will be able to extract vibrational returns from the market indefinitely. All constructs below the pyramidal apex will not require more working capital than originally planned. Entering long positions at successively higher price levels above the apex will eventually entail new capital once upside trend capture constructs like free swing or free styling have used up all the initial capital allocated for this purpose. In some cases, a self-financing situation may develop within the construct, thereby allowing the vibratrader to fund higher long entries by locking in profits at the lower entry levels. Profit is locked in once a stop is placed above the entry price.

A QUICK RECAP

Vibratrading is relatively easy to understand. Below is a quick summary of what you have learned so far in this introductory book.

Let us assume that the vibratrader has sufficient working capital to purchase some shares in an equity ETF at a certain price level. The vibratrader will also have additional capital for more entries should the price of the equity ETF:

1. Fall below the capstone entry level (the entry price of your first entry), or
2. Rise above the capstone entry level.

The vibratrader will be able to predetermine the exact amount of working capital required to initiate long positions at various downbuy levels as prices decline toward ZTL. Some vibratraders may even be able to afford a Martingale downside construct, in which they purchase more shares as prices fall, taking advantage of the PLR, BLR, and MLR associated with these lower price entries while knowing full well the position of the Martingale limit. The vibratrader never executes any downbuys as price declines unless it is part of a pre-calculated move.

The market can in no way whatsoever surprise a vibratrader, except if ZTL is tested.

Should a vibratrader break the fundamental principle of capital boundedness by insisting on a ZTL above \$0, then the vibratrader must be prepared to:

1. Fund the account before this new higher ZTL is tested in order to maintain the downside vibrational constructs.
2. Exit all positions and collapse all constructs at a level slightly above the new ZTL, only to re-enter, or downshift to, a single position at a later date.
3. Exit all positions and collapse all constructs at a level slightly above this new higher ZTL in order to horizontally reinvest the remaining working capital in another equally priced or lower priced instrument.
4. Lose the entire invested working capital.

Since prices cannot test zero in certain instruments or markets, the vibratrader in fact looks forward to large price declines. It is in this scenario that the vibratrader makes the most vibrational returns per interval oscillation, especially with larger positions held at these lower levels.

The vibratrader has five options below the pyramidal apex, namely:

1. To initiate successively lower downbuys, exiting each downbuy in an upsell across one interval, generating purely vibrational returns (Microsiso).
2. To initiate successively lower downbuys, exiting each downbuy in an upsell across two or more intervals, generating more trend-based returns (Macrosiso).
3. To initiate successively higher bounded upbuys when price retraces and triggers a buy limit order and exit each bounded upbuy in an upsell across one interval (Upside Microsiso).
4. To initiate successively higher bounded upbuys when price retraces and triggers a buy limit order and exit each bounded upbuy in an upsell across two or more intervals (Upside Macrosiso).
5. To initiate successively higher unbounded upbuys via buy stop orders and exit each unbounded upbuy in an upsell across two or more intervals (Unbounded Upside Macrosiso).

Remember that "Martingaling" may be applied to the Microsiso and Macrosiso by increasing the scale factor, or by implementing a construct that removes all upside running capital through single or multiple upsells (cost-based construct). The vibratrader can also generate a profit in the process of removing the running capital via profit-based constructs. The returns from Martingale-type constructs are more vibrational than directional as most of the profit drivers (shares, lots, and contracts) are located toward the lower part of the construct, giving the greatest vibrational returns. This is where the vibratrader hopes to dwell as long as possible.

Short positions may also be introduced at any point within the downside vibrational construct, as long as the overall construct is net long or neutral. Net short positions are not permissible under the principle of capital boundedness. These long and short positions allow the vibratrader to extract both long and short oscillational profits, thereby creating a much more efficient profit extraction mechanism. The vibratrader is now extracting profits bidirectionally, rather than merely on the upsells.

Four powerful vibrabidirectional constructs available within the pyramidal structure are:

1. The bounded short pair
2. The big hedge
3. The small hedge
4. Upside short hedge.

Variations are possible when using the big and small hedges as both bidirectional constructs may implement dual or multiple sizing to allow the vibrabidirectional trader to receive trend profits as well as oscillational profits.

The big hedge may also be initiated as a net long mechanism without the linear Microsiso/Macrosiso running along side of it, unlike the version seen in previous chapters (see Figure 16.1).

We see that the big hedge is now directionally exposed to upside trend profit as well as vibrationally exposed to downside short oscillational profits. There are no oscillational returns from the long Microsiso/Macrosiso construct as we have removed it in this version. However, note that trend profits are only possible above \$30, and vibrational returns only below \$30! If price rises significantly above \$30, the shorts may be removed. Every time the shorts are re-established at a higher level, the number of long positions reduces because we must always exit an equal number of longs and shorts to maintain capital boundedness.

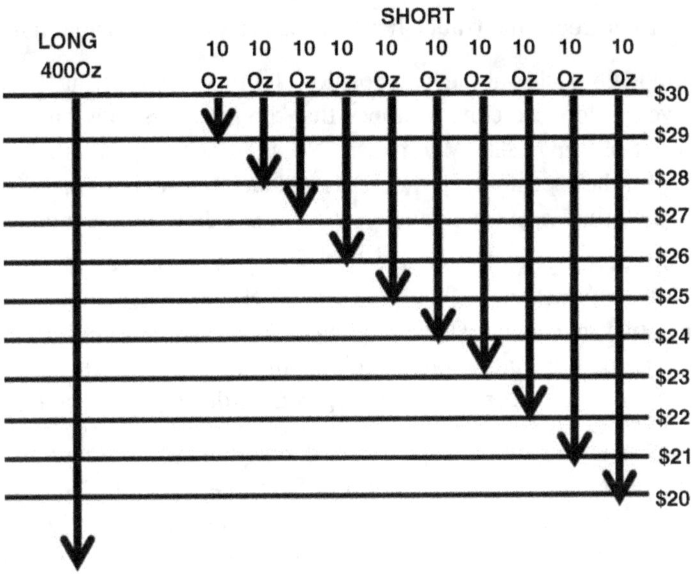

FIGURE 16.1 The Medium Hedge Vibrabidirectional Construct

Another construct, the zero cost hedge, is technically bidirectional as it is comprised of long and short positions. This directionally-neutral construct affords the vibratrader the opportunity to turn back time, so to speak, and only partake in a construct if it is advantageous to do so, without any initial commitment. The shorts are released once price has fallen to an advantageous level (also called loading the matrix), or fallen a significant amount. The zero cost hedge is a useful tool when the vibratrader is uncertain about a market or stock, but still wants to "get in" without any exposure should it prove advantageous.

Now, let us turn our attention to the task of capturing upside trend profits.

Here the vibratrader is seeking to buy high and subsequently sell even higher. In the process of securing trend profits, the vibratrader may also extract vibration returns and accumulate risk-free long positions, if dual sizing and Type 2 exits are used.

A Type 2 exit is a "scale to break-even" stop exit, as described in Chapter 12. Since the main purpose is the capture of trend or directional returns, especially in a buy high–sell higher scenario, the vibratrader is now vibradirectional.

The vibradirectional trader has two options to the upside, namely:

1. To gradually scale out the long positions for profit, re-entering these longs at levels below the exits in hopes that the re-entries may be liquidated again at higher levels (Macrososi). No additional capital is required to scale out these positions if the APD is taken into account. The price level where all the longs are finally scaled out is called the pyramidal apex.
2. To initiate longs to the upside by allocating all or part of the upside working capital to a single price entry level (Free Swinging) or across a range of price levels (Free Styling). Free Swinging and styling are versions of the Macrosimo mechanism. Macrosimo are long upside entries that are exited via sell stop orders during profit taking.

In the free swing, we re-establish long positions without adding capital by freeing up capital through a Type 1 or 2 break-even exit. The free swing vibradirectional trader may even establish smaller long positions for the sole purpose of extracting very short term oscillational returns. This is termed upside scalping. For the first time, no added capital is required to scalp the market. There is also the option to accumulate risk-free long positions during the scalping process via dual sizing.

In free styling, the vibradirectional trader establishes a range of long positions across multiple intervals without any break-even stops. No stop is engaged until the allocated capital is fully utilized. To fund further upside longs, the lowest positions are risk-freed via either a Type 1 or 2 break-even stop.

Besides the larger constructs described above, there are other, simpler versions of upside trend capture like the non-dual sized Upbuy-Upsell-Downbuy-Upsell mechanism.

CHOOSING A VIBRATRADING CONSTRUCT

Your final choice of construct (or combination of constructs) depends on the following:

1. The type of return required
2. The frequency of returns
3. Ease of implementation
4. Level of capital commitment.

The Type of Return

Vibratraders must opt for either a more directional or a more vibrational construct. Constructs more directional in nature include all Macrosimo-type constructs like the free swing, free styling, and the non-dual sized mechanism. More vibrational constructs include all Microsiso and Macrosiso constructs, like the standard linear (SF = 1) downbuy construct and the non-linear Martingale downbuy construct (SF > 1). The more bidirectional constructs are the big and small hedges and Macrosiso vibrahedging I and II.

The Frequency of Returns

Vibratraders looking for more rapid oscillational returns must choose either a more volatile market or reduced interval sizing of the constructs. Unless Interval Reduction Technique II is used, the cost will escalate very rapidly with finer intervals. Another way to increase the frequency of returns is to establish a pure vibrabidirectional construct that extracts both long and short vibrational returns, like those listed above.

Ease of Implementation

Vibratraders may opt for constructs that are easy to establish and maintain. The simplest upside construct to implement is the Macrososi. The simplest downside construct would be the Microsiso, all with a simple scale factor of one.

Level of Capital Commitment

Vibratraders may choose to trade with minimal commitment of capital. On the next page is an example which combines the two least capital-intensive constructs for extracting returns from both upside and downside moves. To the upside we enter a non-dual sized vibrational construct with an Upbuy-Upsell-Downbuy-Upsell configuration. The vibratrader will be exposed to no more than the minimum lotsize permissible on the trading platform at any one time, which in the following example is 10 ounces (see Figure 16.2).

We use the small hedge to the downside, because it is the lightest downside construct available. It is even lighter than a regular downbuy

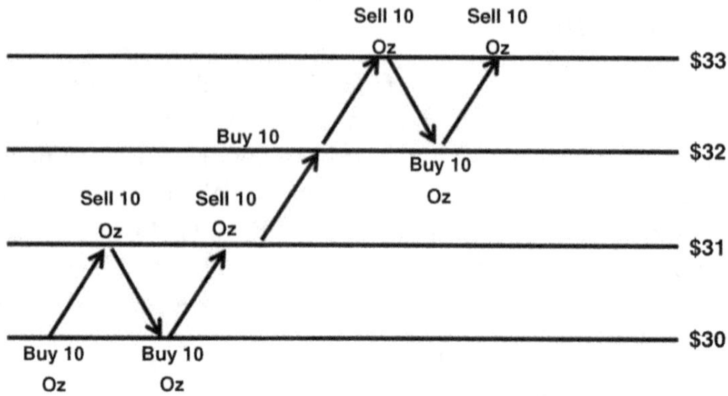

FIGURE 16.2 Non-Dual Sized Vibrational Construct

construct with a scale factor of one, since the short trend profits (if dual sized) and short vibrational profits help reduce the equity drawdown during a decline. The small hedge morphs seamlessly into the non-dual sized construct and vice versa to form the least capital-intensive construct for capturing vibrational returns regardless of price direction (see Figure 16.3).

Another construct that is relatively easy to implement is the Macrososi, but only at a very low price entry level. For example, if you purchased 7,000

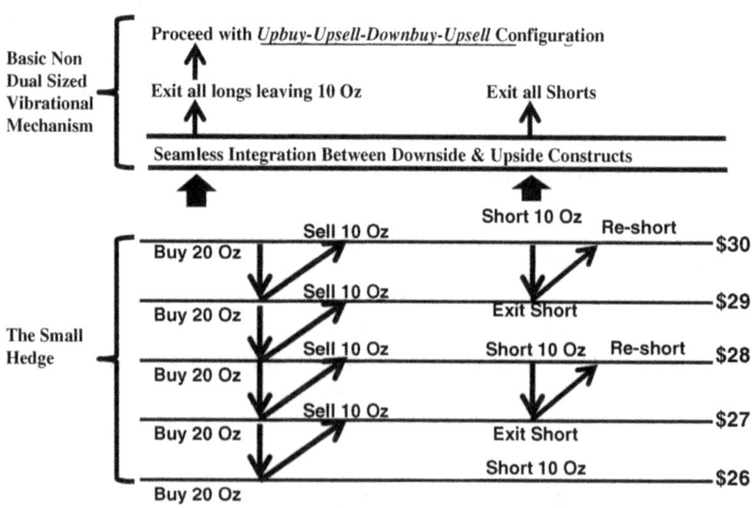

FIGURE 16.3 Seamless Intergration Between Constructs

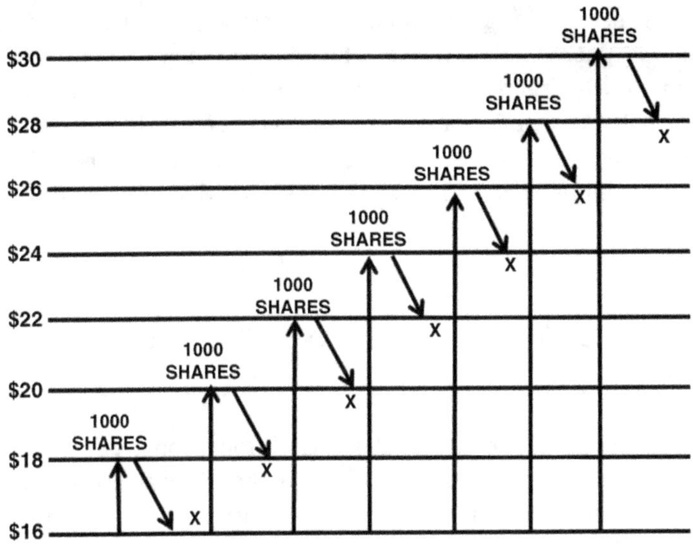

FIGURE 16.4 The Macrosiso Scale Out in Action

shares of an ETF stock, you could design a construct to scale out of a certain number of shares per fixed interval to the upside. The shares sold are then bought back at exactly one interval below their exit, ready for the next upsell to take place. This establishes a vibrational construct. The set-up is essentially an inverted big hedge—the longs are scaled out to the upside instead. There is no need to hedge the construct as there are no shorts contained within it (see Figure 16.4).

Initially, 7,000 shares are purchased at $16. The first 1,000 shares are exited at $18 for a $2,000 profit. Should price retrace down to $16, the 1,000 shares are bought back, ready for another round of oscillational profit. The re-entries are marked "X" in Figure 16.4. As price rises, another 1,000 shares are exited at $20 for a profit, only to be re-purchased at $18 should price retest that level. Again, this sets up conditions for more vibrational profit if price oscillates around the $18 to $20 levels.

In fact, the vibratrader could design a Macrososi construct to span from the very lowest entries to any price level desired. This creates a very wide upside vibrational construct that runs from near ZTL right to the apex. The height of the apex is only limited by the minimum lotsize or sharesize permissible on the trading platform and the maximum disposable capital available. As seen in earlier chapters, the Macrososi will transform into the Microsiso or Macrosiso upon re-establishing the long entries.

Besides the bidirectional constructs, vibratrading offers the participant an opportunity to consistently extract non-directional returns from inactive, consolidating, and even flatlining markets via the use of riskless short options embedded within the Microsiso and Macrosiso constructs.

THE IMPORTANCE OF A BALANCED PYRAMIDAL STRUCTURE

The pyramidal structure contains all the vibrational constructs within it. As long as it is capital bounded, a construct is part of the structure. Remember that you may place different constructs at various price levels for the traded instrument or market.

Generally, a pyramidal structure should have three sections:

1. A trend capture construct at the upper end
2. A downside vibrational construct
3. A protective construct at the lowest levels near ZTL.

The protective construct is typically a Martingale construct, as they are fairly capital intensive and so are better placed at the lowest price levels, where PLR, BLR, and therefore MLR, will yield the greatest returns.

The Pyramidal Structure and Trend Capture Constructs

As with all pyramidal structures, trend capture constructs are not designed to work across the entire price range. These constructs do not usually work to the downside unless there is an element of bidirectionality within the construct.

Hence it is important that the vibratrader always have at least two constructs in operation: one for trend capture and the other a standard downside vibrational construct, like the Microsiso or Macrosiso. This dual construct set-up will allow the vibratrader to capture upside trend as well as downside vibrational gains should price access the lower levels of the construct.

CONCLUSION

Above all, vibratrading teaches the participant one very important lesson: **Always enter the market well prepared, with all the groundwork completed and all scenarios accounted for in detail**. This detailed planning is only possible because vibratrading is price independent. As a vibratrader, you meticulously plan every entry and exit within the construct. Once completed, there is virtually nothing that the market can do to catch you off guard, short of a truly catastrophic event. The vibratrader has no real preference as to whether the market rises or falls, as long as it keeps rising and falling.

I have endeavored to build a methodology whereby market participants will have the tools to make a living from the markets without the need to constantly maintain a positive expectancy. That must be the greatest advantage of vibratrading: there are no loss exits, and hence the usual reference to the ubiquitous win/loss percentage is inapplicable. No more endless back-testing and optimization to find out if the methodology is profitable, but rather optimizing your scaling parameters to get a greater profit. No more time spent figuring out the safest location to position a stop—instead you determine the best location to lock in profit with a stop. No more paralyzing analysis trying to figure out where the market will move next, but rather a more objective analysis in figuring out what you would do next should the market move in that direction.

Index

www.ingramcontent.com/pod-product-compliance
Lightning Source LLC
LaVergne TN
LVHW011515200926
848484LV00001B/44

* 9 7 8 0 4 7 0 8 2 8 7 4 8 *